D0838855

Book Three

Developing Number Concepts

Place Value, Multiplication, and Division

Kathy Richardson

Math Perspectives

Managing Editors: Alan MacDonell, Catherine Anderson
Developmental Editors: Harriet Slonim, Beverly Cory
Editorial Advisor: Deborah Kitchens

Production/Manufacturing Director: Janet Yearian
Production/Manufacturing Coordinator: Joan Lee
Design Director: Phyllis Aycock
Design Manager: Jeff Kelly
Text Design: Don Taka
Cover Design: Lynda Banks
Cover Illustration: Christine Benjamin
Text Illustrations: Linda Starr
Composition: Claire Flaherty

Many of the designations used by manufacturers and sellers to distinguish their products are claimed as trademarks. Where those designations appear in this book, and the publisher was aware of a trademark claim, the designations have been printed in initial caps.

Copyright © 1999 by Math Perspectives Teacher Development Center. All rights reserved. No part of this book may be reproduced or transmitted in any form or by means, electronic or mechanical, including photocopying, recording, or by any information storage and retrieval system, without permission in writing from the publisher. For information regarding permission(s), write to Rights and Permissions Department.

This book was previously published by: Pearson Education, Inc.

ISBN 978-0-9848381-6-5
Printed in the United States of America
11 12 13 06 05 04

www.mathperspectives.com

Preface

In the years since I wrote *Developing Number Concepts Using Unifix® Cubes*, I have come to believe even more strongly in my teaching approach for ensuring children's success in mathematics. Whether it is the research on how the brain works or the new learning theories that are being refined or developed, the information we have continues to validate my belief that children need meaningful experiences that engage their thinking. We teachers are coming to a renewed appreciation of how vital our role is to the learning process. While we recognize that the child must do the learning, we also recognize that we are an important part of that learning. When we know our children well and fully understand what we are trying to teach them, we can ask questions, provide experiences, and set up situations that maximize their learning.

As we continue to grow as a profession, we teachers see that there are some things that withstand the test of time. I have found that one of these things is this approach to the teaching of mathematics to young children and how it affects the way children learn. It seemed to me to be a worthy endeavor, therefore, to work to make the information that I set forth in my first book more accessible and easier to use in this series. If you have worked with my first book, you will be familiar with many of the activities in these next books. You will see how the activities have been updated by the use of other manipulatives—in addition to Unifix Cubes—and by the grouping together of all levels of any particular activity. In these books you will find questions to guide your observations of the children at work and more help in meeting the wide variety of children's needs. And, you will find realistic classroom scenes that will help make my teaching approach come to life.

Despite these changes from my original book, the actual teaching techniques remain the same with the exception of the treatment of two topics—double-digit addition and double-digit subtraction. If you have been using my first book, you may want to take a special look at how my thinking about the best way to teach these topics has evolved. And so, I hope that if you are one of the many teachers who found your teaching to be helped by the old "Brown Book" (as some have called it) you will find this new series to be even more helpful.

KR

Acknowledgments

No one ever writes a book alone. The words that appear on the final pages come from the myriad of experiences that an author brings to the task. There are more people than I can name who led me to and helped me through the writing of this series. To all the teachers and teacher leaders I have worked with over the years, and to all the children who let me enter their lives and learn from them, I give thanks. And to everyone in my large and wonderful family, ever patient with my preoccupation with my work, I give my gratitude and love.

There are some people who were particularly important in helping me get this project done and to whom I would like to express my appreciation. The old cliché "I couldn't have done it without you" is once again true. Special thanks to Deborah Kitchens and Janann Roodzant. The phrase "countless hours" comes to mind when I think of all the time they dedicated to helping me. And thanks to Jody Walmsley, Patti Boyle, Marilyn Smith, Kathy McGrath, and Barb Escandon—who kept me in touch with real kids and real classrooms—to Ruth Parker who continues to stretch my thinking, and to Linda Gregg—who kept me moving and wouldn't let me give up. Special thanks to Linda Starr for bringing these books to life through her delightful illustrations.

Some thoughts never dim with the passing of time. In memory of Mary Baratta-Lorton, who is always present in my work and in my heart.

Credits

Unifix® Cube is a registered trademark of Philograph Publications, Ltd.

Snap™ Cube is a trademark of the Cuisenaire® Company of America.

Baratta-Lorton, Mary (1976). *Mathematics Their Way.* Menlo Park, CA: Addison-Wesley Publishing Company, Inc.

Richardson, Kathy (2015). *Assessing Math Concepts.* Bellingham, WA: Math Perspectives Teacher Development Center.

Richardson, Kathy (1990). *A Look at Children's Thinking.* (Video II) *Assessing Math Concepts.* Bellingham, WA: Lummi Bay Publishing Co.

D'Nealian® Handwriting is a registered trademark of Donald Neal Thurber. Used by permission of Addison Wesley Educational Publishers, Inc.

Table of Contents

Table of Contents

CHAPTER ONE (continued)

Table of Contents

Table of Contents

Blackline Masters

The blackline masters numbered for use with this book are listed below. (For a complete listing of the blackline masters used in the *Developing Number Concepts* series see pages 211–212.)

Developing Number Concepts—the Series

Developing Number Concepts is a series of books designed to help young children develop important foundational mathematics concepts.

Each of the three books in the series includes cohesive and organized sets of experiences focused on particular mathematical ideas. Every concept is developed both through teacher-directed and independent activities. Because children learn at different rates the activities are "expandable" and, therefore, meet a range of needs. Questions that guide teachers' observations of children as they work and learn help in the assessment of children's ongoing progress.

Book One

Chapter 1: ***Beginning Number Concepts***
Chapter 2: ***Pattern***
Chapter 3: ***The Concepts of More and Less***

Book Two

Chapter 1: ***Interpreting and Symbolizing Addition and Subtraction***
Chapter 2: ***Internalizing Number Combinations to 10***
Chapter 3: ***Developing Strategies for Adding and Subtracting***

Book Three

Chapter 1: ***Place Value***
Chapter 2: ***Beginning Multiplication***
Chapter 3: ***Beginning Division***

The Planning Guide for Developing Number Concepts accompanies the series. It is for the use of teachers of kindergarten through grade three and teachers of multi-grade classes. It includes comprehensive year-long teaching plans along with classroom management ideas.

How the Books Are Organized

Each chapter of *Books One, Two,* and *Three* includes the following.

- **What You Need to Know About...**

This section provides the teacher with background information on the featured math concept and a summary of ways in which to teach the concept.

- **Chapter Overview**

A brief overview of the chapter follows. It offers pertinent information on how the math concept should be taught to children at each grade level, kindergarten through grade three, and to children with special needs.

- **Goals for Children's Learning**

This section lists the mathematics concepts, ideas, and skills that the children will learn as they work with the activities.

- **Analyzing and Assessing Children's Needs**

Questions to guide teachers' observations and a discussion of how the activities can be used to meet a range of needs are included. The questions are geared to help teachers determine if the tasks that children are working with are appropriate and are meeting their needs.

- **Classroom Scenes**

Realistic classroom scenes that deal with the major math concepts covered in the chapter help bring the activities to life as they model ways in which the teacher can work.

- **About the Activities**

Included here is a brief discussion about the purpose of the activities along with information about materials preparation.

- **Teacher-Directed Activities and Independent Activities**

A great variety of both teacher-directed and independent and/or partner activities are included for each math concept. This gives teachers many different ways to meet children's needs while it gives children many different ways to learn about a particular concept.

- **Blackline Masters**

Blackline masters, used both for materials preparation and as children's worksheets, appear at the end of each book.

Goals and Purposes

Children's first experiences with numbers will influence the way they deal with mathematics for the rest of their lives. Children will benefit from and will be able to build on their early experiences if they learn mathematics in ways that make sense to them. These books are for teachers who want to make mathematics understandable to children and who want to help children build a mathematics foundation that will serve them in the years to come.

Guiding Principles

The approach presented here is based on certain principles of how children learn. It will be most effective if presented in light of these principles, some of which follow.

Children develop an understanding of concepts through experiences with real things rather than symbols. Teachers and parents have known for generations that children learn from real-life experiences. Such wisdom grew from observing children over time. It has been validated by the latest research on how the brain works and by current educational theories on how children learn. The activities presented in this series engage children in exploring, discovering, and interacting with mathematical concepts in ways that get them to think critically. The approach goes even further as it offers a wealth of activities that provide children with ongoing practice that will enable them to develop facility and fluency with mathematical ideas. In the past, children may have practiced number concepts by working solely with workbook exercises. The activities in *Developing Number Concepts* provide the practice children need but in ways that are more engaging and meaningful.

Teachers can support the development of understanding by presenting planned and focused experiences and by interacting with the children as they work. Children must come to their own understanding of the concepts and gain competence and facility for themselves. Teachers play an important role in supporting concept development by planning mathematical experiences for children that help them confront, interact with, and practice particular mathematical ideas. When teachers have children work with sets of related activities that meet a range of needs, they can observe all the children at work and then interact with them individually in ways that enhance children's learning.

For children to be engaged by a particular mathematical task, they need to be on the edge of their understanding or level of competence. Children will find tasks engaging if the experiences meet their needs for developing new understandings or for developing confidence or competence with a new skill. Developing competence is a prime motivator for young children. They will naturally choose to practice a seemingly simple task over and over again until they no longer find it challenging.

When working independently, children should be allowed to choose from a group of related tasks. Children do not all work and learn at the same pace. They do not all stay interested and focused for the same amounts of time. When teachers provide children with a choice of tasks, they can each select the one they think is the most interesting and valuable. They are more likely to be productively engaged when they are in control of—and are, therefore, responsible for—having made that particular choice. Children should be free to choose to do any particular task over and over again. They get the most from a task when they are free to come back to it as many times as they think is necessary.

The most powerful learning experiences have value in being repeated. Many enthusiastic teachers believe that in order to provide the best math program possible, they must continually present new and different activities. However, this may not be the best way for children to develop an understanding of the concepts that are the most important for them to learn. Learning how to do an activity is just the beginning. Not until children fully understand how to do a task are they ready to learn from it. When children are encouraged to work with familiar tasks over time, they will get the full benefit from the experience. The teachers' focus can then shift from making sure the children know what to do to using the experience to help children deepen conceptual understandings.

The Manipulatives

In order for number concepts to be meaningful to them, children must experience numbers as they occur in the real world. Children need to count, compare, combine, and take apart numbers using a variety of manipulatives. Their work with manipulatives helps them discover mathematical relationships and build visual images. There are many different manipulatives that can be used as effective models of mathematical ideas and relationships. Some of these are commercially available and others can be gathered by and/or made by teachers. The following manipulatives are suggested for use with the activities in this series. Classroom quantities of each are specified.

Connecting, or Interlocking, Cubes

Many of the activities require the use of ¾-inch connecting cubes. These manipulatives can be linked to form "trains" or "towers." Approximately 1,000–2,000 connecting cubes will be needed. They can be found under the following brand names:

Unifix® Cubes
Snap™ Cubes

Counters

A great many activities rely on the use of counters. A variety of manipulatives should be used as counters, thus enabling children to experience the same activity with different models. The following manipulatives can be used effectively as counters:

Unifix® Cubes or **Snap™ Cubes** (used individually)
Color Tiles (400–800)
Wooden Cubes (500)
Collections (15–20) Each collection should consist of 60–100 small items of one kind, such as buttons, shells, screws, pebbles, and bread tags.

Additional Manipulatives

Some activities require specific manipulatives, such as the following:

Toothpicks, flat (3 or 4 boxes)
Pattern Blocks (3 sets)
Beans (2 lb) and portion cups, available in restaurant-supply stores (for place-value activities)

The following charts identify mathematics concepts that children need to know and understand. They also list the Book Three **(3:)** activities—teacher-directed and independent—that can be used to support children's learning of the concepts. Some activities meet a variety of needs and so are listed in several places. Refer to the section entitled "Questions to Guide Your Observations," in each Chapter Overview, to help you determine those needs.

Chapter 1: Place Value

Section A: Understanding Regrouping—The Process and the Patterns

If your children need...	Teacher-Directed Activities	Independent Activities
practice in forming groups and counting groups: The following activities help children focus on the processes of grouping and regrouping as they work with groups of four, five, and six objects. (Working with groups of less than ten helps children to focus on the grouping process and to make the appropriate generalizations.)	3: 1–1 Introducing the Plus-One and Minus-One Games 3: 1–2 The Grouping Games with Groups of Other Sizes 3: 1–3 Plus or Minus Any Number 3: 1–4 Regrouping Beyond Two Places	
to record the patterns that emerge from forming groups: Children discover patterns that emerge as they work with small groups. Later, this will form the basis of their understanding of base-ten number patterns.	3: 1–5 Number Patterns in the Plus-One and Minus-One Games 3: 1–7 Introducing Number Patterns in a Matrix	3: 1–6 Recording the Plus-One and Minus-One Patterns, Ext. 3: 1–8 Recording the Patterns in a Matrix
to form groups of tens and to record the patterns that result: The children will learn how to group numbers by tens, identifying the patterns that result and identifying two-digit numbers as groups of tens and leftovers.	3: 1–9 Introducing Grouping by Tens 3: 1–12 Patterns on the 00–99 Chart	3: 1–10 Writing Base-Ten Patterns on a Strip 3: 1–11 Creating a 00–99 Chart
to connect number patterns to various patterns using manipulatives: Children explore growing patterns using manipulatives. Then they label these growing patterns with numbers.	*(The following activity is from Book One.)* 1: 2–15 Exploring Growing Patterns 3: 1–13 Naming Patterns with Colors 3: 1–14 Analyzing Growing Patterns 3: 1–15 Finding the Number Patterns in Growing Patterns	*(The following two activities are from Book One.)* 1: 2–17 Growing-Pattern Task Cards 1: 2–18 Creating Growing Patterns 3: 1–19 Number Patterns in Growing Patterns

Chapter 1 (continued)

Section A: Understanding Regrouping—The Process and the Patterns

If your children need...	Teacher-Directed Activities	Independent Activities
to search for patterns in base ten, to understand the patterns formed by numbers to 100, and to become familiar with the 00–99 chart:	3: 1–10 Writing Base-Ten Patterns on a Strip 3: 1–16 Introducing Pattern Searches	3: 1–18 Grab and Add 3: 1–19 Number Patterns in Growing Patterns 3: 1–20 Margie's Grid Pictures 3: 1–21 Looking for Patterns on the 00–99 Chart 3: 1–22 The 00–99 Chart Puzzles 3: 1–23 Searching-for-Patterns Station

Section B: Developing a Sense of Quantities to 100 and Beyond

If your children need...	Teacher-Directed Activities	Independent Activities
to develop a sense of quantities for numbers to 100: The children practice partitioning large numbers in order to develop flexibility in working with numbers as well as to develop an understanding of conservation of large numbers. They make estimates and then determine actual amounts and learn to organize numbers into groups of tens and ones for ease in counting.	3: 1–24 Rearrange It: Arranging Loose Counters into Tens and Ones 3: 1–25 Rearrange It: Breaking Up Trains into Tens and Ones 3: 1–26 Rearrange It: Finding All the Ways 3: 1–27 Rearrange It: How Many Cubes? (10–20) 3: 1–27 Rearrange It: How Many Cubes? (numbers beyond 20) 3: 1–28 Rearrange It: Breaking Up Tens 3: 1–29 Build it Fast 3: 1–30 Give-and-Take with Tens and Ones 3: 1–31 Think About the Symbols	3: 1–32 Lots of Lines, Level 1 3: 1–33 Paper Shapes, Level 1 3: 1–34 Yarn, Level 1 3: 1–35 Yarn Shapes, Level 1 3: 1–36 Containers, Level 1 3: 1–37 Cover It Up, Level 1 3: 1–38 Measuring Things in the Room, Level 1 3: 1–39 Measuring Myself, Level 1 3: 1–41 Making Trails, Level 1 3: 1–43 Race to 100 3: 1–44 Race to Zero
to compare quantities to determine which of two quantities is more and which is less:		3: 1–32 Lots of Lines, Level 2 3: 1–33 Paper Shapes, Level 2 3: 1–34 Yarn, Level 2 3: 1–35 Yarn Shapes, Ext. 3: 1–36 Containers, Level 2 3: 1–38 Measuring Things in the Room, Level 2 3: 1–39 Measuring Myself, Level 2 3: 1–40 Comparing Myself 3: 1–42 Building Stacks

Chapter 1 (continued)

Section B: Developing a Sense of Quantities to 100 and Beyond

If your children need...	Teacher-Directed Activities	Independent Activities
to compare quantities to determine how many more one quantity is than another:		3: 1–32 Lots of Lines, Level 3 3: 1–33 Paper Shapes, Level 3 3: 1–34 Yarn, Level 3 3: 1–36 Containers, Level 3 3: 1–38 Measuring Things in the Room, Level 3 3: 1–39 Measuring Myself, Level 3
to develop a sense of quantities for numbers beyond 100:		3: 1–32 Lots of Lines, Ext. 3: 1–33 Paper Shapes, Ext. 3: 1–34 Yarn, Ext. 3: 1–35 Yarn Shapes, Ext. 3: 1–36 Containers, Ext. 3: 1–41 Making Trails, Ext.

Section C: Addition and Subtraction of Two-Digit Numbers

If your children need...	Teacher-Directed Activities	Independent Activities
to practice solving problems in a variety of ways: The children use manipulatives to learn to interpret addition and subtraction problems and develop their own strategies for determining how many.	3: 1–45 Addition and Subtraction of Two-Digit Numbers 3: 1–46 Story Problems 3: 1–47 Figure It Out	3: 1–48 Partner Add-It 3: 1–49 Partner Take-Away 3: 1–50 Roll and Add 3: 1–51 Roll and Subtract 3: 1–52 Add 'Em Up: Lots of Lines 3: 1–53 Add 'Em Up: Paper Shapes 3: 1–54 Add 'Em Up: Measuring Things in the Room 3: 1–55 Add 'Em Up: Yarn 3: 1–56 Add 'Em Up: Yarn Shapes 3: 1–57 Add 'Em Up: Containers 3: 1–58 Add 'Em Up: Cover It Up 3: 1–59 Solving Story Problems

Chapter 2: Beginning Multiplication

If your children need...	Teacher-Directed Activities	Independent Activities
to develop an understanding of multiplication by searching for and counting groups: The first step in understanding multiplication is to develop the idea of counting equal groups. These activities focus children's attention on multiplication situations as they can occur in the real world. Children are not yet required to write multiplication equations.	3: 2–1 Looking for Equal Groups in the Real World	3: 2–11 How Many Cups? 3: 2–12 How Many Groups? 3: 2–13 How Many Rows? 3: 2–14 How Many Towers?
practice in acting out multiplication stories: These experiences help children build an understanding of the process of multiplication and provide a meaningful basis for later work with symbols.	3: 2–2 Acting Out Multiplication Stories: Using Real Objects 3: 2–3 Acting Out Multiplication Stories: Using Counters	
practice in interpreting the language of multiplication using physical models: Children need to model multiplication in a variety of ways. These include working with cube trains or towers and stacks, rows, piles, and groups of counters.	3: 2–4 Building Models of Multiplication Problems	
to look for relationships when working with related multiplication problems: This should be an ongoing experience for children as they explore what happens when they look for relationships. Repeat this activity on occasion throughout children's work with multiplication.	3: 2–5 Building Related Models	

Chapter 2 (continued)

If your children need...	Teacher-Directed Activities	Independent Activities
practice in reading and interpreting multiplication equations: Children need to connect their multiplication experiences and actions to symbolic representations. These activities should be presented to the children before you ask them to write equations.	3: 2–6 Modeling the Recording of Multiplication Experiences 3: 2–7 Introducing the Multiplication Sign 3: 2–8 Interpreting Symbols	3: 2–15 Counting Boards: Multiplication, Level 1
practice in writing multiplication equations to describe a problem or situation: Children write their own equations to label various situations. These activities provide children with practice in reading, interpreting, and writing multiplication equations.	3: 2–10 Learning To Write the Multiplication Sign	3: 2–15 Counting Boards: Multiplication, Levels 2 and 3 3: 2–16 Problems for Partners: Multiplication 3: 2–17 Roll and Multiply 3: 2–18 Discovering Patterns: Cupfuls, Level 1 3: 2–19 Discovering Patterns: Buildings, Level 1 3: 2–20 Discovering Patterns: Number Shapes, Level 1 3: 2–23 Lots of Rectangles 3: 2–24 Shape Puzzles: Multiplication
to discover the patterns that occur when working with multiplication: Looking for patterns and seeing the same patterns occur over and over again will help children learn multiplication equations.	*(The following activity, from Chapter One, may be used either to introduce pattern searches or to review them.)* 3: 1–16 Introducing Pattern Searches	3: 2–18 Discovering Patterns: Cupfuls, Level 2 3: 2–19 Discovering Patterns: Buildings, Level 2 3: 2–20 Discovering Patterns: Number Shapes, Level 2 3: 2–21 Pattern Search: Multiplication
practice in writing story problems to go with an equation: Writing stories adds another dimension to interpreting equations. This provides children with a way to permanently record their ideas and helps them to connect multiplication with the world outside their classroom.	3: 2–9 Acting Out Stories To Go with Multiplication Problems	3: 2–22 Writing Stories To Go with Multiplication Problems

Chapter 3: Beginning Division

If your children need...	Teacher-Directed Activities	Independent Activities
to develop an understanding of division as the sharing, or partitioning, of equal groups: Children need these kinds of experiences to build an understanding of the process of division and to provide a meaningful basis for later work with symbols.	3: 3–1 Acting Out Division Stories: Using Real Objects 3: 3–2 Acting Out Division Stories: Using Counters	
practice in interpreting the language of division using physical models: Children need to model division in a variety of ways. These include breaking up cube trains or towers into equal stacks, rows, piles, or groups.	3: 3–3 Building Models of Division Problems	
to use division to look for relationships: The activity should be repeated on occasion as the children explore these relationships.	3: 3–4 Odds and Evens	
to learn to read and interpret division equations: Children need to connect their division experiences and actions to corresponding symbolic representations.	3: 3–5 Modeling the Recording of Division Experiences 3: 3–6 Interpreting Symbols	3: 3–9 Counting Boards: Division, Level 1
to learn to write division equations to describe a problem or situation: Children need to connect their division experiences and actions to symbolic representations. This should be presented to the children before they are asked to write equations independently.	3: 3–7 Learning To Write the Division Sign	

Chapter 3 (continued)

If your children need...	Teacher-Directed Activities	Independent Activities
practice in reading, interpreting, and writing division equations:		**3:** 3–9 Counting Boards: Division, Levels 2 and 3 **3:** 3–10 Number Shapes: Division **3:** 3–11 Making Rows **3:** 3–12 Problems for Partners: Division **3:** 3–13 Cups of Cubes **3:** 3–14 How Many Buildings? **3:** 3–15 Creation Cards for Division
practice in writing story problems to go with equations: Writing stories adds another dimension to interpreting equations. This provides children with a way to permanently record their ideas.	**3:** 3–6 Interpreting Symbols, Ext.	*(The following activity is from Book Two.)* **2:** 1–10 Writing Stories To Go with Equations
practice in relating multiplication and division:	**3:** 3–8 Multiplication and Division Together: Story Problems	

1

When textbook or curriculum objectives are:

- Naming digits in the ones, tens, and hundreds places
- Writing numbers to 100 (and beyond)
- Comparing two 2-digit numbers
- Skip counting
- Adding and subtracting two 2-digit numbers

Then you want to teach

Place Value

SECTION A

Understanding Regrouping: The Process and the Patterns

SECTION B

Developing a Sense of Quantities to 100 and Beyond

SECTION C

Addition and Subtraction of Two-Digit Numbers

What You Need to Know About Place Value

Place-value concepts are not easy for young children to grasp. Many children, exposed to these ideas too soon, too quickly, and too abstractly, remain uncertain and confused about them throughout elementary school. Year after year, children ask their teachers questions like, "Do we have to borrow?" and "Is this when we carry?" Children diligently memorize steps and rules for getting answers, often forgetting or misapplying them. The teacher's job in the primary grades is to build a foundation so that children will be able to make sense of place value and determine for themselves whether or not they are getting reasonable answers when solving problems. The following are several ideas that children must deal with as they come to terms with place-value concepts.

When we count large numbers, we keep forming new groups based on tens.

The most basic concept confronting children is that our number system is based on forming groups of ten. That is, when we have ten units, we group them into one group of ten. When we have ten groups of ten, we group them into one group of one hundred. When we have ten groups of one hundred, we group them into one group of one thousand, and so on.

We count groups as though they were single objects.

For children who have been counting single objects, counting groups requires a shift in thinking. Children's first counting experiences involve an understanding of one-to-one correspondence. They learn that each count can be matched to one object. But when dealing with numbers beyond ten, they are required to count groups as though they were individual objects.

We can look at one amount as being made up of many different groups.

Children need to understand that ten ones is the same as one group of ten, and that 100 ones is the same as one group of 100 or ten groups of ten. Posing the question, "How many tens are there in 34?" assumes that the child can conceive of ten objects as one entity. Posing the question, "How many hundreds are there in 346?" assumes that the child can conceive of 100 objects as one entity (all the while remembering that each hundred is also ten groups of ten).

A numeral (digit) can stand for different amounts, depending on where it is written in the number.

Another key idea the children must learn is that a particular numeral can stand for many different amounts, depending on its position, or *place,* in a number. Imagine for a moment what it would be like if every number we could think of had to be written with a unique symbol—such a thought is overwhelming.

1 2 3 4 5 6 7 8 9

We can all be grateful that our number system requires us to learn only ten different numerals that we can put together to write *any* number, no matter how large. This system works because the value of each numeral changes according to the size of the group it stands for. The size of the group is indicated by the numeral's position in the number.

As wonderful as this system is, learning it is quite a big step. When children have just recently learned that "7" stands for one more than six objects, it is asking a lot to expect them to understand that "7" can also stand for other amounts, depending on its position in a number. For children who still don't see the difference between 7 and ٦, or *saw* and *was,* or who are still learning to tell their left hand from their right hand, place value can be quite a mystery.

What makes this concept even more complicated is that although one number can be represented correctly in several ways, it is incorrect if written in slightly different ways. For example, *seventy* can be represented as 7 tens and 0 ones, or as 70, but not as 7 or 07. Similarly, *seventy-eight* can be represented as 78, 70 + 8, 7 tens and 8 ones—even as 8 ones and 7 tens—but not as 87 or 708. To many children, these differences are very subtle.

What seems to be difficult and potentially so confusing to children is possible to deal with because of the patterns in the number system which make it so orderly and predictable. Once we have discovered the patterns in the number system, the task of writing numerals to 100 (and beyond) is simplified enormously. We encounter the same sequence of numerals (0, 1, 2, 3, 4, 5, 6, 7, 8, 9) over and over again. Children do not see the patterns automatically, so they need experiences that allow them to discover these patterns.

Our number system is based on patterns that help us see relationships.

In Book One, we discussed the idea of conservation of number. Children come to understand the notion of conservation of number when they realize that the *quantity* of a group of objects does not change when those objects are moved closer together, or farther apart, or are hidden or rearranged in some way. Even when children have come to understand conservation of *small* numbers, some children as old as 8 still may not have developed an understanding of conservation of *large* numbers. To such children, it is not obvious that 26 cubes grouped into 2 tens with 6 left over is still 26. The simple action of rearranging, or regrouping, may seem like an operation such as addition or subtraction, and so children may think that after regrouping they end up with a number that is different from the one they started with.

Children who are 7 and 8 years old are still developing their understanding of conservation of large numbers.

Addition and Subtraction For many years, children have learned procedures for getting answers to two-digit addition and subtraction problems in spite of the fact that they may not understand the underlying math concepts. To really understand these procedures, children must be able to understand not only adding and subtracting but also complex place-value concepts. That is, they must understand that numbers are constructed by organizing quantities into groups of tens and ones, they must understand how numerals change in value depending on their position in a number, and they must be able to take apart two-digit numbers and put them back together again.

When children are asked to learn procedures before they understand the underlying concepts, they are forced to memorize steps that they have no way of understanding.

Asking children to memorize steps for solving problems actually creates barriers to the development of understanding. However, when we give children experiences that help build the basic concepts they need and allow them to make sense of the procedures in their own ways, they become able to add and subtract in ways that are more effective and certainly more meaningful to them. When allowed to make their own sense of addition and subtraction, children become better able to use these processes to solve problems than when they are expected to learn the teacher's way of determining sums and differences. All the activities in Section C of this chapter are based on this premise.

The Use of Models to Teach Place-Value Concepts The use of manipulative materials such as connecting cubes or beans and cups can help children in dealing with place-value concepts. The unique qualities of connecting cubes can make them especially effective tools. The children can physically join cubes into a single unit so that they see ten single objects become one object. Joining these ten units to form one "train" does not cause the units to disappear. The children can still count and check, if necessary, to see how many units they have. This activity is a beautiful model of the idea that the quantity of ten can be both one ten and ten ones at the same time.

If a child starts with fourteen loose cubes and snaps ten of them together into a train, all that has changed is the *arrangement* of the cubes. The child still has the same fourteen cubes that he or she started with. Connecting cubes have the potential to help children develop an understanding of conservation of number in a way that is not possible with an abacus or with blocks of various sizes designed to represent hundreds, tens, and ones. Beans and small paper cups can also be used to rearrange numbers into groups of hundreds, tens, and ones without changing the actual objects being worked with.

Models of the base-ten number system have different levels of abstraction that may not be meaningful to children.

Some math manipulatives used to teach place-value concepts require children to trade ten counters for one of a different kind of counter. When children are just beginning to figure out what regrouping is all about, trading can be an extra step that clouds their understanding of the concept. The original counters they started with are gone, and there is no way for children to verify that what they now have is equivalent to what they started with—they simply have to take the teacher's word for it. There is a time when trading many objects for one object is appropriate, especially when dealing with very large numbers. However, manipulatives such as base-ten blocks or colored chips should not be used too soon.

Place-value concepts are not easy to learn, especially for young children. Children's understanding will develop and broaden over a long period of time. The activities in this chapter, designed to help children begin to understand, will serve to support them as they explore these ideas and grow in their understanding.

Teaching and Learning About Place Value

The activities in this chapter help children develop an understanding of place-value concepts including forming and counting groups, recognizing patterns in the number system, organizing groups into tens and ones, and adding and subtracting two-digit numbers. The chapter is organized into three sections. Each section deals with important concepts that children must understand if they are to make sense of our number system and learn to add and subtract large numbers in a meaningful way.

In each section, you will introduce the concepts to the children through teacher-directed whole-class activities. You will provide follow-up in the form of independent activities that give children the opportunity to consider and apply these ideas for themselves.

Section A: Understanding Regrouping—The Process and the Patterns
The whole class is introduced to the underlying concepts of place value through a set of activities called the grouping games, through which children practice forming and counting groups. They record and examine the patterns that emerge from this process. The children then build on this experience as they work with groups of ten, interpreting numbers as "tens and leftovers," and searching for patterns in base-ten numbers.

Section B: Developing a Sense of Quantities to 100 Children develop a sense of quantities to 100 and come to recognize the value of organizing numbers into groups of tens and ones. They learn to take numbers apart in a variety of ways, a process that develops flexibility with numbers and an understanding of conservation of large numbers.

Section C: Addition and Subtraction of Two-Digit Numbers Children learn to interpret addition and subtraction problems with manipulative materials and develop their own strategies for determining sums and differences.

Using the Chapter

Your use of this chapter will vary according to the needs of your children. The "Meeting the Needs of Your Children" charts in the introduction to this book and in the *Planning Guide* that accompanies this series offer detailed information that can help you plan how to use the chapter's activities. The following are general suggestions for using the activities with different groups of children.

Kindergarten The place-value concepts in this chapter are inappropriate for use in kindergarten. While it may be of some value to children to experience organizing objects into groups of tens and ones in situations like counting pumpkin seeds, the children would be generally learning to use the place-value language without really understanding the underlying concepts.

First Grade Although many first-grade children can count to 100 and are intrigued by looking for number patterns, most first graders will not be able to understand place-value concepts. In order to understand place value, they must be able to think of ten in two ways at the same time, both as one ten and as ten ones. (The difficulty that this kind of thinking presents to young children has a parallel in a more common situation: at a certain stage of thinking, young children can't accept the idea that their mother is both their mother and their grandmother's daughter at the same time.) First graders' work with large numbers is, for the most part, focused on counting. Such work prepares them for later work with place-value concepts. If you introduce place-value concepts to your first graders, be aware that many will not understand the concepts or the language they are learning to use.

Second Grade Place-value concepts are the core of the second-grade mathematics program. Second-grade children will explore the ideas of forming and counting groups and discover the patterns that emerge from this process. They will have opportunities to discover the patterns in the place-value system and will see how these patterns occur in their environment. They will engage in many different activities that require them to work with tens and ones and that help them to develop a sense of quantity of numbers to 100. They will also work with addition and subtraction of two-digit numbers and will develop their own strategies for determining the sums and differences.

Third Grade The amount of work that third-grade children need to do with the place-value concepts presented in this chapter will depend in large part on their previous experiences. Most third-grade children will benefit from working with these concepts in the beginning of the year. Extend the activities so that children are working with numbers beyond 100 and are adding and subtracting three-digit numbers.

Children with Special Needs Because place-value concepts are difficult for children to understand, we often ask children simply to memorize procedures rather than giving them sufficient time to develop understanding. Do not rush children through these experiences. They will be better served in the long run if they are allowed to make sense of what they are learning. Provide them with lots of experiences organizing groups into tens and ones. Give the children a great deal of practice counting groups of tens using real things. Give them many opportunities to interpret and build models of two-digit numbers using beans and cups or connecting cubes.

SECTION A

Understanding Regrouping: The Process and the Patterns

The activities in this section help children understand the underlying concepts of place value so that they will be able to work effectively with our base-ten number system. Children form groups in a systematic way, counting and recording the groups and the leftovers and then identifying the number patterns that emerge.

Instead of beginning to work with groups of ten, the children will work forming smaller groups of four, five, and six. This is because working with groups of less than ten enables children to focus more easily on the process of regrouping, since they are forming groups more often than if they were working with groups of ten. Later, when the children work with tens and hundreds, the regrouping process will already be clear to them.

The children also learn to count groups and to write the numbers that stand for those groups in the appropriate places. They search for the patterns that emerge when forming groups in a regular fashion. The repetitions in the patterns that emerge when working with groups of four, five, and six are shorter and thus more easily recognizable than similar patterns in base ten. The experiences identifying these shorter patterns will help the children later when they begin to look for patterns in base ten.

Once the foundation has been laid using smaller groups of numbers, the children apply what they have learned as they work with groups of ten. The patterns in the base-ten number system then become the focus of study as the children discover and work with a variety of patterns in the numbers to 100 and beyond.

Goals for Children's Learning* *(Section A)*

Goals

Given a variety of situations, the children will:

- Form and count groups
- Interpret numbers as groups and leftovers
- Interpret numbers as tens and ones (or hundreds, tens, and ones) using models
- Identify patterns in sequences of numbers
- Find numbers with ease on the 00–99 chart
- Count and write numbers to 100 and beyond

* Adapted from *How Do We Know They're Learning? Assessing Math Concepts.*

Analyzing and Assessing Children's Needs

Children need to understand much more about place value than just how to name numbers in the hundreds, tens, and ones places. They need to know how to interpret these numbers and model them with manipulatives. It is not enough that they can write numbers to 100. They need to be able to see the patterns inherent in the number system and recognize similar patterns as they occur in the environment. You will be able to find out what your children understand about place value by observing them as they work. The following questions will guide your observations.

Questions to Guide Your Observations*

Questions

Forming and Counting Groups

Observe the children when they are working with groups of four, five, and six and again when they are working with groups of ten.

- Are the children able to form and count groups?
- Do they anticipate when they will have enough to form a group or do they forget to form groups when necessary?
- Can the children move easily from forming groups of one size to forming groups of another size?
- Are the children able to count and record the groups and leftovers?

Working with Two-Digit Numbers as Tens and Ones

- When asked to interpret a number written as tens and ones, such as 52, can the children do this accurately?
- Do they describe the numbers in the tens place as a number of tens using the language "50" or "5 tens," rather than calling it just "5"?
- Can they show tens and ones appropriately with models?

Number Patterns

- Do the children easily recognize the patterns that emerge or do they need help?
- Do they use the patterns to identify any errors that occur?
- Can the children see any similarities or other relationships between the patterns?

* Adapted from *How Do We Know They're Learning? Assessing Math Concepts.*

Patterns in Base Ten

- Can the children write numerals to 100 using patterns to help them?
- Can they fill in blanks in the 00–99 chart in a variety of ways? That is, can they fill in the surrounding numbers appropriately when given only a few numbers on the chart?

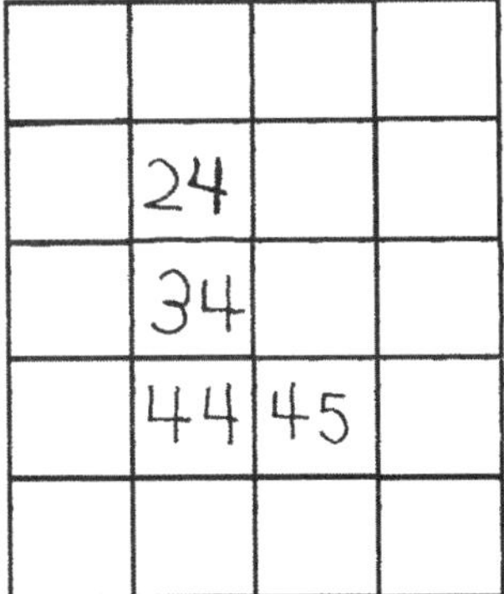

- What patterns do children notice on the 00–99 chart?
- When working with the 00–99 chart, do they use a sense of pattern to find the numbers on the chart, or do they need to count or search to find particular numbers?
- Do they discover and can they describe number patterns when adding by twos, fives, tens, and so forth?

Labeling Real-World Patterns with Numbers

- Can the children use numbers to describe a variety of patterns that occur in the real world?
- Can they identify and sort the number patterns that occur in various situations?

Meeting the Range of Needs

As you introduce children to the process of regrouping through the whole-class, teacher-directed lessons, move through the activities at a pace that is appropriate for most of the children. The goal is to get the children working independently, each at his or her own pace, as quickly as possible. You can then provide extra support to any children who need more time with the introductory activities. Once the children are searching independently for patterns, they will be able to work at their own levels. More information on meeting the range of needs is included with particular activities.

A Classroom Scene

Mrs. Williams' second-grade class is well into the search for number patterns. A few weeks ago they explored adding by twos and discovered the 0, 2, 4, 6, 8, ... pattern.

The children wanted to be able to refer to this pattern in an easy way, so as a class they decided to call it the *red* pattern. One of the children colored the pattern red on the 00–99 chart, and her record was posted in the room.

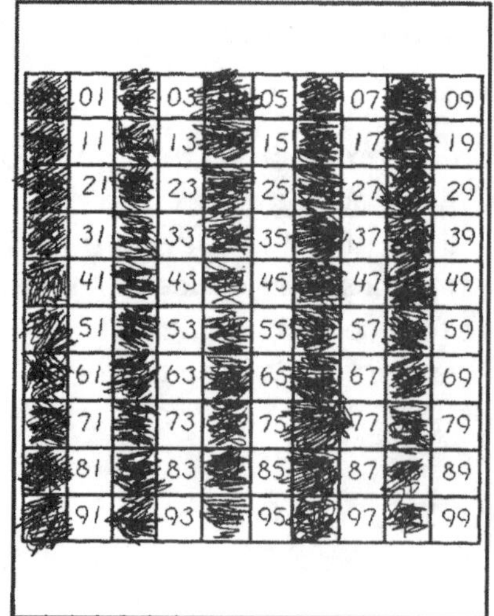

Whole-Class and Independent Work: *Discovering and Sorting Number Patterns*

A few days later, Mrs. Williams' class worked together to find a number pattern for counting fingers. As the children came up to the front of the room one at a time, the class counted fingers by fives while Mrs. Williams showed them how to write the pattern in a T-table.

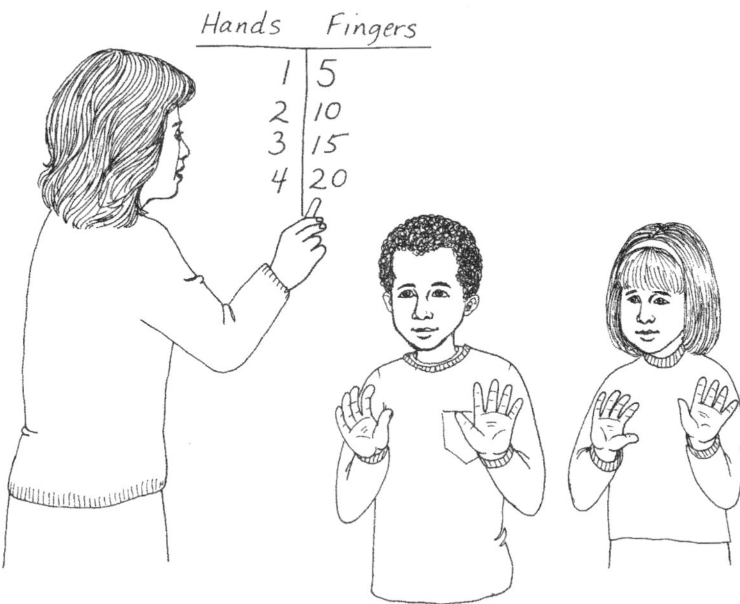

The class decided to call this pattern the *blue* pattern. Dimitri colored the pattern blue on a 0–99 chart, and his record of the fives was posted next to the red pattern. The children individually made a record of this experiment as the first page in their pattern book.

	01	02	03	04		06	07	08	09
	11	12	13	14		16	17	18	19
	21	22	23	24		26	27	28	29
	31	32	33	34		36	37	38	39
	41	42	43	44		46	47	48	49
	51	52	53	54		56	57	58	59
	61	62	63	64		66	67	68	69
	71	72	73	74		76	77	78	79
	81	82	83	84		86	87	88	89
	91	92	93	94		96	97	98	99

00
05
10
15
20
25
30

Later that same week, the class discovered the pattern for adding by tens and labeled that pattern *purple*. The children had previously discovered a number pattern that "grew" by three with each count. They labeled it the *orange* pattern.

Today, Mrs. Williams has the children working at lots of different stations to see what patterns they can discover (Searching-for-Patterns Station, activity 1–23). At one table, the children are using toothpicks to make triangles. At another table, the children can see what pattern they find when they count eyes.

There are yellow pattern blocks (hexagons) and paper pattern-block hexagons at one table, red pattern blocks (trapezoids) at another table, and orange pattern blocks (squares) at still another table. The children are going to investigate the patterns for the number of sides of a shape.

Each time the children discover a pattern, they are to decide if it is like one of the patterns they have already labeled and posted on the wall or if it is a new pattern. They will share their findings at the end of the work time.

These experiences are preparing the children for a pattern search that will continue throughout the year. Once the common number patterns have been identified and labeled with a color, the children will continue to look for those patterns in many different situations.

About the Activities in Section A

In other chapters of this book, the teacher-directed and independent activities are presented separately. This section of Chapter One departs from that format. In some lessons, the entire class follows each step of the teacher's instructions. In other lessons, the whole class works independently on a particular task. After the sequence of lessons that prepares the children to form groups and to record and look for patterns, the children can then work with several different pattern stations and explore a variety of patterns.

The Grouping Games

The Process of Regrouping Start with a series of activities called the grouping games, first presenting whole-class activities that teach the children to count groups as though they were single objects. It is important to include all your students in the introduction of the place-value grouping games, even though you may feel that not all of them are ready at the same time or that not all of them should move through the place-value activities at the same pace. All the children should be involved in choosing nonsense words that name the groups so that later they will know where the words came from.

Discovering Number Patterns Once the children are comfortable with the grouping games, the whole class should learn to look for the patterns that emerge when the steps of the grouping games are recorded (activity 1–5). After the recording of patterns has been introduced, the children will be able to work independently to discover patterns.

Patterns in Base Ten The grouping games provide a foundation for understanding base ten. In this series of activities, the children work with a variety of number patterns, including adding by twos, fives, and tens. Through these activities, children become familiar with many patterns in our number system.

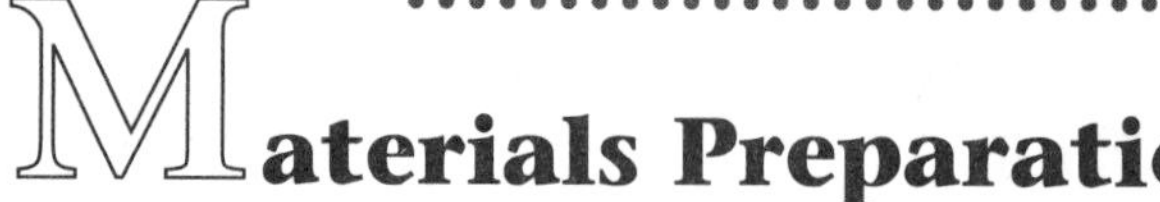

Materials Preparation

Every attempt has been made to keep the materials preparation to a minimum. Generally, the materials needed are described in each activity. One material used repeatedly is the place-value board. Children use the place-value boards to organize their counters into tens and ones.

Place-Value Boards

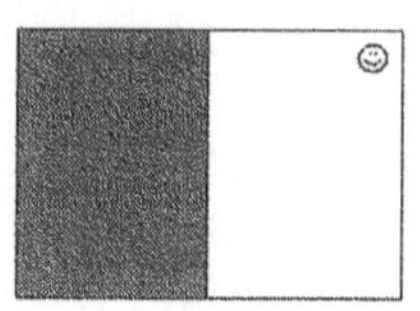

Staple or glue a 6" × 9" piece of colored paper onto the left side of a 9" × 12" piece of white paper or tagboard. Draw a tiny happy face in the upper right-hand corner so that the children are sure to position the board properly by seeing that the happy face is right-side-up.

Teacher-Directed and Independent Activities

About the Grouping Games

The grouping games teach the children a particular process for forming and counting groups. You will go through this process step by step with the children, using either connecting cubes or beans and cups with the place-value boards, until you are sure that the children understand the procedure and can apply it to whatever size group they are working with.

When you first introduce the grouping games, the children work with groups of four. Later they will work with groups of five and six. As a class, they pick a nonsense word to name each group. Naming a group gives it an identity and helps the children think of it as one entity. While introducing a nonsense word into mathematics may seem unnecessary, referring to a group of objects in this way is actually less confusing to children than referring to it with a number like *four* or *five*.

1-1 Introducing the Plus-One and Minus-One Games* ..... Whole-Class Activity

Materials: Connecting cubes (sorted by color) or beans and cups ▪ Place-value boards (1 per child). (See Materials Preparation, p. 14.)

The Plus-One Game

The first task is choosing a word to describe the group you are going to work with. Say, for example:

Today we are going to play a counting game. We will be counting groups of cubes (or beans). In this game, we are going to make groups of four, but we can't say the word "four." We need another name for these groups, so we have to make up a new word that means "four." The word can't mean anything else. Who has an idea?

I have an idea. "Zib."

(This is just an example of a nonsense word. Your students will invent their own word.)

OK, "zib." I will write it on a chart so we won't forget.

* Based on *Mathematics Their Way,* "Counting Game: Concept Development Stage," p. 276.

To start the Plus-One game, each child should have a place-value board and a supply of connecting cubes.

When we count zibs, we will use our place-value boards to help us organize our counting. The green side of the board will be where we put the zibs, and the white side will be where we put loose cubes when we don't have enough for a zib.

When I say "Plus one," I want you to put one cube on the white side of your board. Okay, "Plus one."

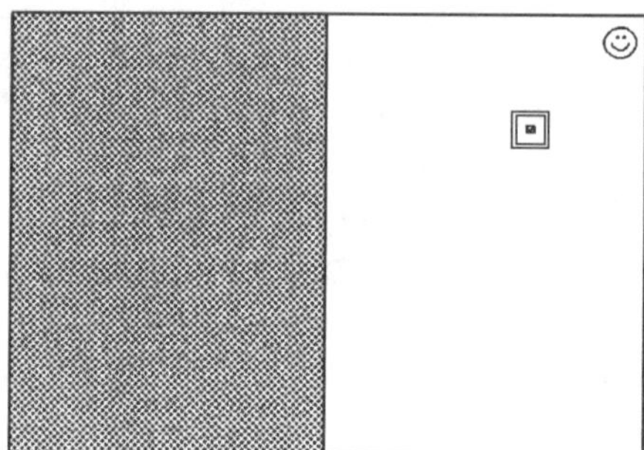

Next, ask the children to say what they have on the board by telling how many zibs and how many ones there are. (You want the children to tell how many of each at each step so that later they will be able to write each step and look for the patterns that emerge.)

How many do you have on your board now?

Zero zibs and one.

Initially, say what is on the board, for example, "zero zibs and one," *with* the children until the children pick up on what they are supposed to say and are able to say it without you.

Keep on saying "Plus one" and asking "How many?" ("zero zibs and two, zero zibs and three...") until there are four cubes, or one zib, on the white side of the board. Then have the children snap the cubes together and put the newly formed zib on the green side of the board.

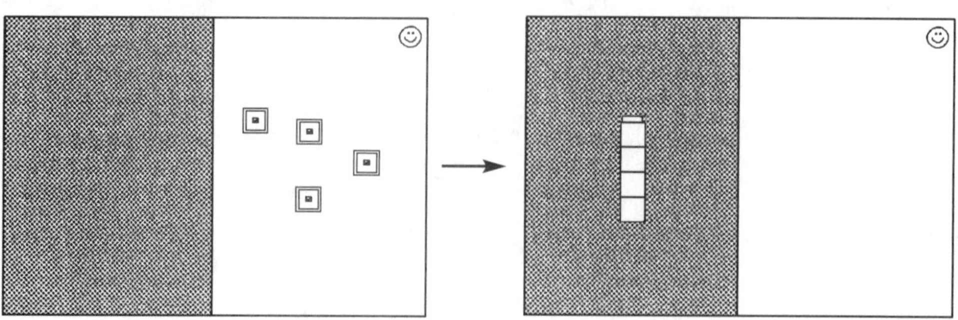

The first time you work with this activity, you will need to emphasize when it is time to make a group (a zib, in this example). After the children get the idea, they will be anticipating when they have enough to make a group and will be saying things like, "We have another zib!"

Continue the game, saying "Plus one" and asking "How many?" until the children each have three zibs and three on their board.

NOTE: In the beginning, it will be easier for the children if they do not have to regroup beyond two places. If you were to continue to say "Plus one" after the children have three zibs and three on their boards, the children would then have another zib. If they put that zib on the green side of the board, then they would have four zibs. Whenever the children reach four zibs, they must regroup the four zibs into another group, placing the zibs into a container to the left of the board to form a new group called a *big zib*.

Using Cubes

Big Zib

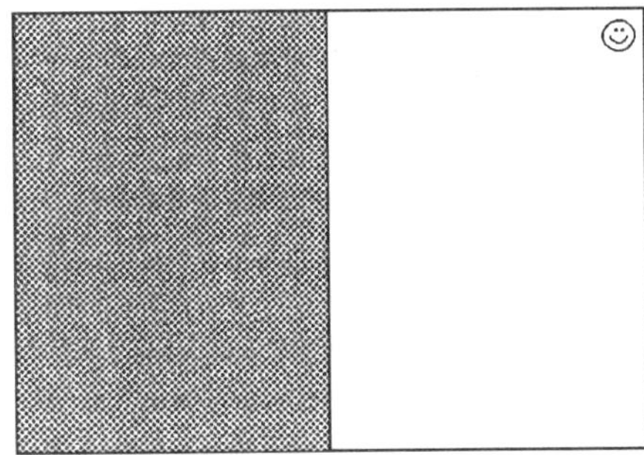

Using Beans

Big Zib

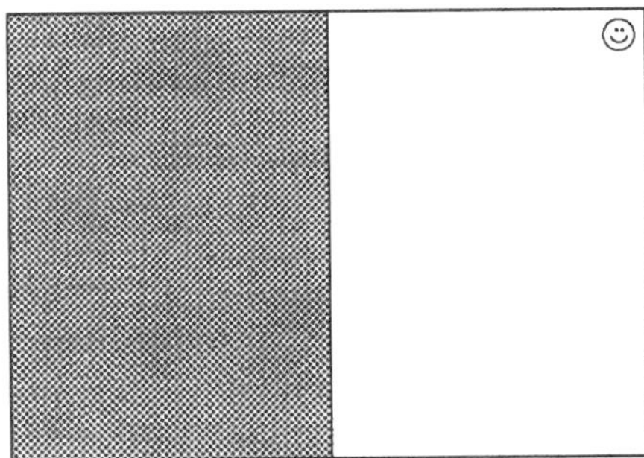

To avoid dealing with big zibs in these first lessons, stop saying "Plus one" when three zibs and three is reached. See activity 1–4 for a description of what to do when you are ready to extend the regrouping beyond two places.

The Minus-One Game

When three zibs and three is reached, tell the children that you are going to start saying "Minus one."

Minus one. How many do we have now?

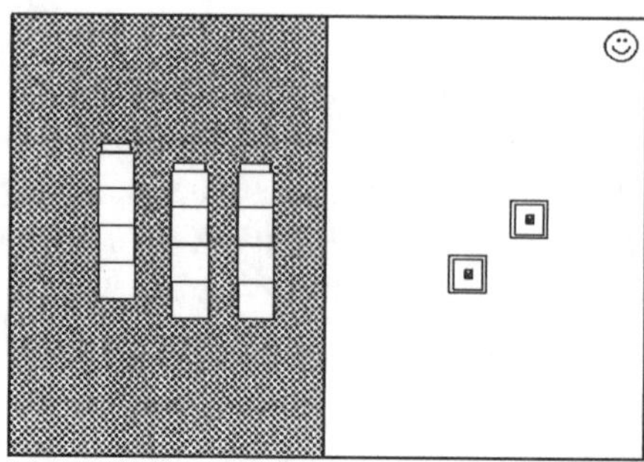

Three zibs and two.

Continue saying "Minus one" until you reach three zibs and zero.

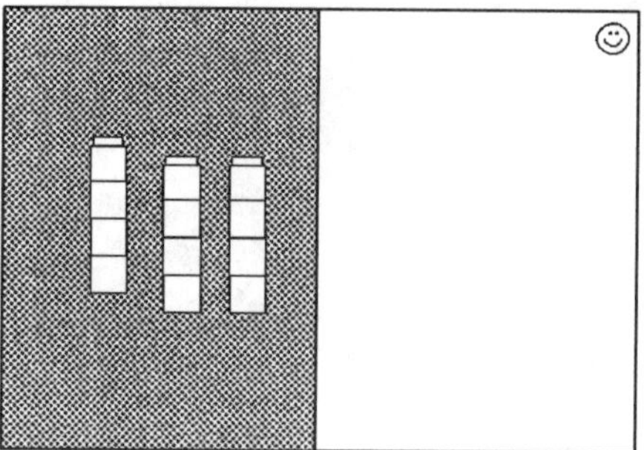

At this point, talk to the children about what they can do to take one cube away. Some will naturally break up one zib into four loose cubes from which they can take one away. Others will break off one cube and leave the rest of the zib on the green side of the board. Help the children see that three joined cubes no longer form a zib, so they can't remain on the green side. For example:

If you take one cube off a zib, is it still a zib?

> **No. There's only three.**

Only zibs belong on the green side—so what shall we do?

> **Put three cubes on the white side.**
>
> **Yeah, but first we have to break them up.**

How many zibs do we have now?

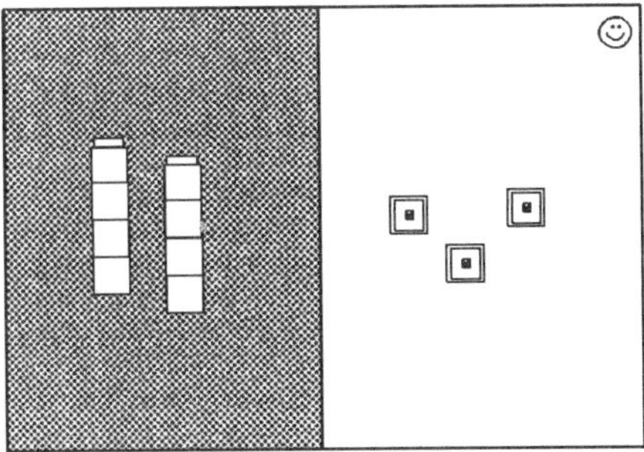

Two zibs and three.

Minus one. How many?

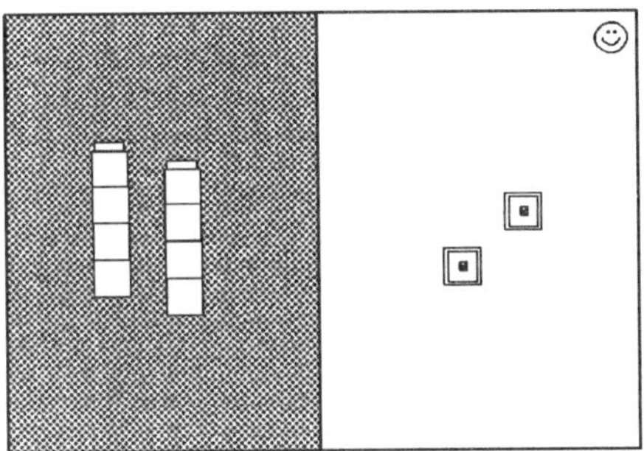

Two zibs and two.

Continue saying "Minus one" until you reach zero zibs and zero.

Repeat this activity on succeeding days. Vary it sometimes by using beans and cups with the place-value boards.

You do not want the children to think that the grouping games are played only with groups of four, so after two to five more lessons, begin playing with groups of other sizes as described in the following activity.

1–2 The Grouping Games with Groups of Other Sizes

Whole-Class Activity

Materials: Connecting cubes (sorted by color) or beans and cups ▪ Place-value boards (1 per child)

In order for children to develop an understanding of the concept of regrouping, they will need experience in making groups of various sizes. After two to five lessons in grouping by fours, introduce grouping by fives. Have the children choose a nonsense word to stand for a group of five and then add it to your classroom chart. For example:

We are going to play another counting game. In this game we are going to make groups of five, but we can't say "five." We need a new word. Who has an idea?

Glom.

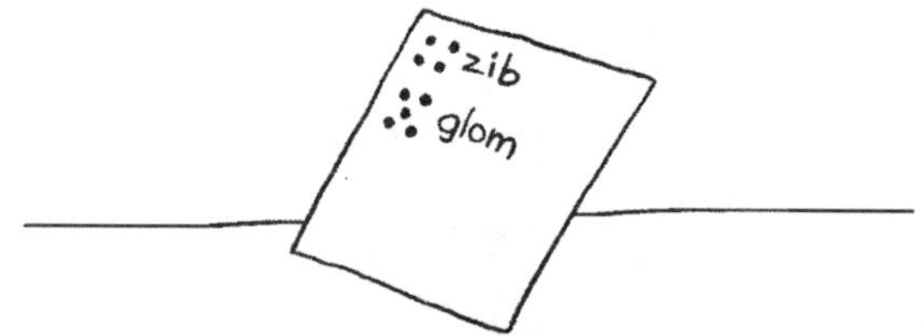

As necessary, explain that the word "zib" was a label for a group of four and so it cannot be used when counting gloms. Point out that children can see if they have a glom by counting "One, two, three, four, glom."

Play the "glom" game with the place-value boards and connecting cubes. Start by playing the Plus-One game as you count up to four gloms and four. Then play the Minus-One game as you count back to zero gloms and zero.

Encourage the children to anticipate the need to make a group by occasionally asking:

How many more do we need to make a glom?

On succeeding days, do the activity again using beans and cups.

After a few days, repeat the game with groups of six. Choose a new word for groups of six, and add it to the class chart. Using the place-value boards and connecting cubes or beans and cups, play the Plus-One game counting up to five groups and five, then play the Minus-One game counting back to zero groups and zero.

After the children are comfortable with groups of four, five, and six, spend a few days working with the groups, in random order, until you feel that the children are able to regroup easily. Be sure to play both games.

Although it is not necessary to explore groups of numbers other than four, five, and six in order for children to understand regrouping, many children become intrigued with the process and ask to explore larger groups. Give those children a chance to work with groups of seven, eight, nine, or of any other size.

1–3 Plus or Minus Any Number .... Whole-Class Activity

Materials: Connecting cubes (sorted by color) or beans and cups ▪ Place-value boards (1 per child)

This activity is similar to the basic grouping games, but here, instead of adding or subtracting one cube each time, the children add or subtract various numbers of cubes according to what you call out. This encourages thinking and flexibility, as the children will be forming groups and "leftovers" as appropriate, rather than on cue. For example:

We are working with zibs today. Listen carefully because I'm going to say to add and to subtract different numbers. Sometimes I'll say "plus" and sometimes I'll say "minus."

Plus two. How many?

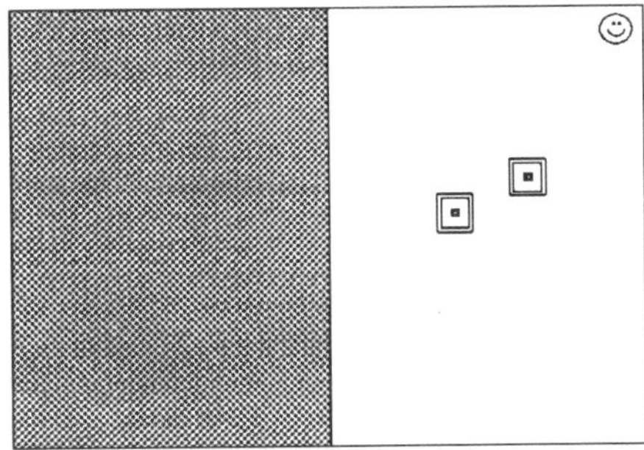

Zero zibs and two.

Plus three.

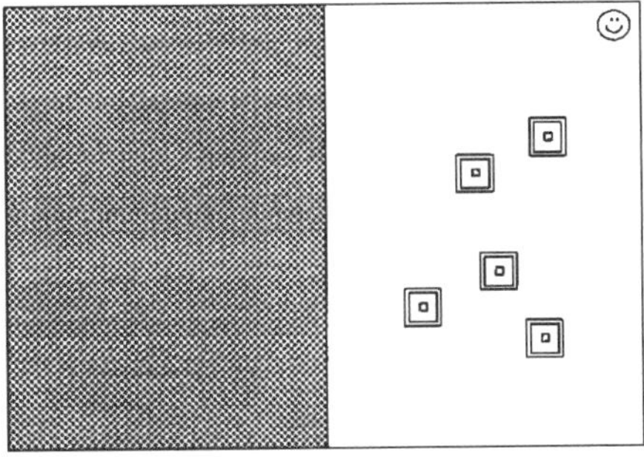

We have enough for a zib. But there's one left over.

How many?

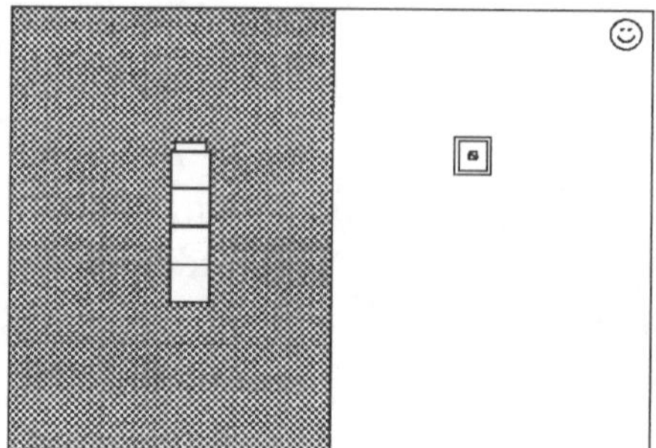

One zib and one.

Plus two. How many?

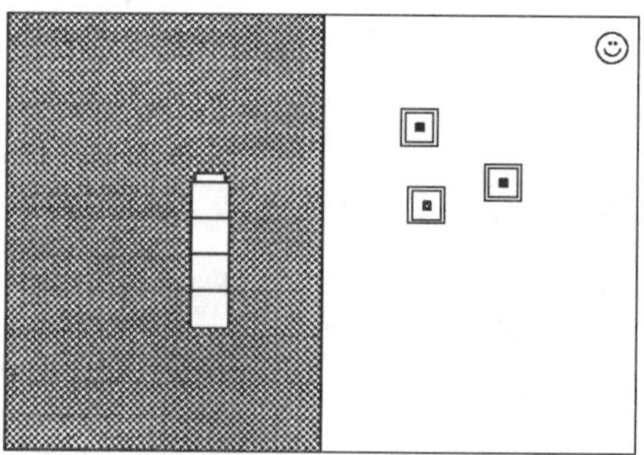

One zib and three.

Plus three.

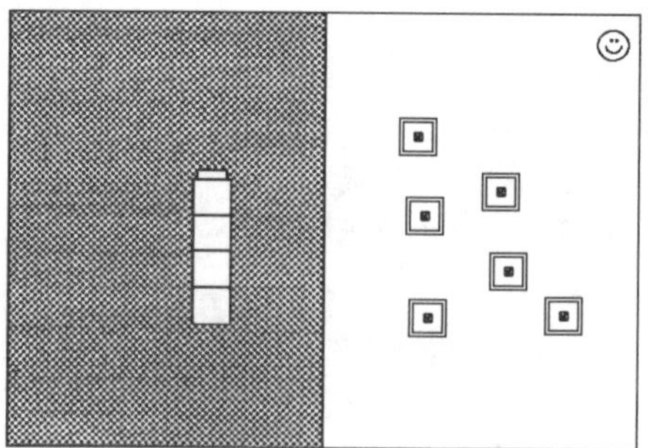

We have another zib.

Now how many?

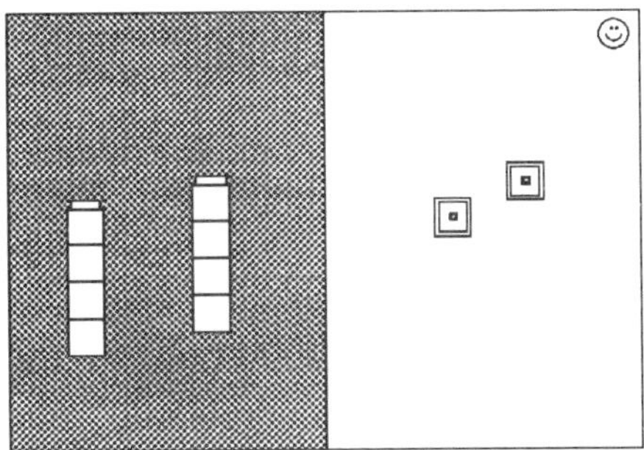

Two zibs and two.

Minus three. How many?

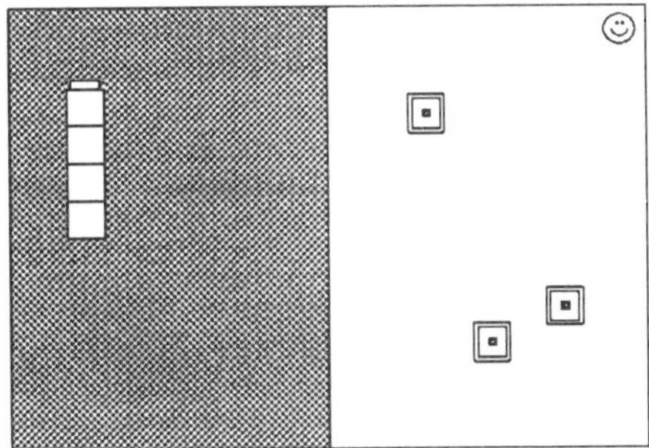

One zib and three.

Repeat this activity on succeeding days. Vary it sometimes by using beans and cups with the place-value boards.

1-4 Regrouping Beyond Two Places

Small-Group Activity

Materials: Connecting cubes (sorted by color) or beans and cups • Place-value boards (1 per child) • Margarine tubs

As children work with groups of any size, they need to see that regrouping is an ongoing process. Regrouping must occur every time the number being worked with is reached. For example, if children are working with groups of five, they need to understand the following.

- When you have five ones, you need to form a group of five.
- When you have five groups of five, you need to form a *big five.*
- When you have five big fives, you need to form a *super five,* and so on.

Model the regrouping process while working with small groups of three and four children. The children already know that when working with groups of four, if they have four cubes on the white side of the board, the cubes must be regrouped into one zib. You will need to explain the next step:

***When you have four zibs, they must be regrouped into one* big zib.**

Add a sheet of paper to the left of the place-value board and put a margarine tub on it to hold the big zib.

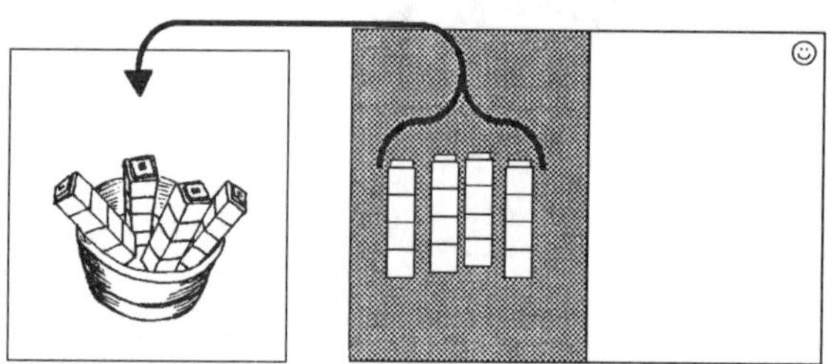

***When you have four big zibs, they must be regrouped into one* super zib.**

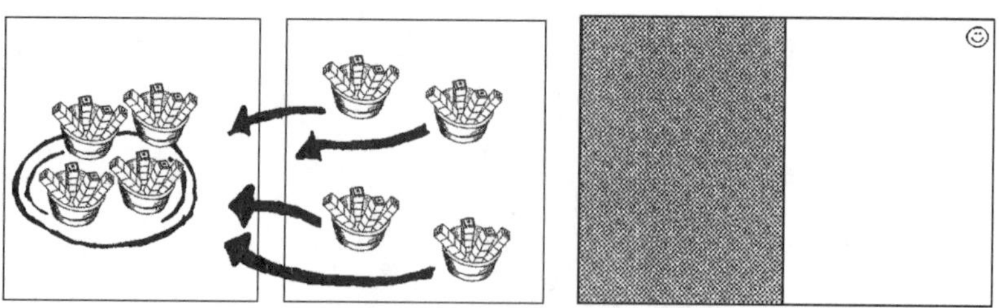

***When you have four super zibs, they must be regrouped into one* gigantic zib.**

Add another sheet of paper and place a plate on it to hold the gigantic zib. When you work with small numbers like these, the need to regroup occurs often, and so the pattern for the process quickly becomes apparent to the children. Later, when they work with tens, hundreds, and thousands, the regrouping process will be already clear to them.

1–5 Number Patterns in the Plus-One and Minus-One Games* ....Whole-Class Activity

Materials: Connecting cubes (sorted by color) or beans and cups ▪ Place-value boards (1 per child) ▪ Long strip of paper for recording the patterns (a roll of paper towels, shelf paper, or butcher paper)
Preparation: Draw a long vertical line on the long strip of paper to create two columns. Shade the left column with a green crayon or marker.

Once the children can make groups easily, you can help them see the number patterns they form. At this stage, continue working with groups of four, five, and six, as these number sequences are shorter and the patterns they form are easier to see than they would be for larger groups of numbers.

Plus-One Number Patterns

Play the Plus-One grouping game in the usual way, but this time record each step of the pattern as the children report it. For example:

We are going to play the "glom" game today. When you tell me the number of cubes you have on your board, I am going to write the numbers down on this long strip of paper.

How many cubes are on your boards right now?

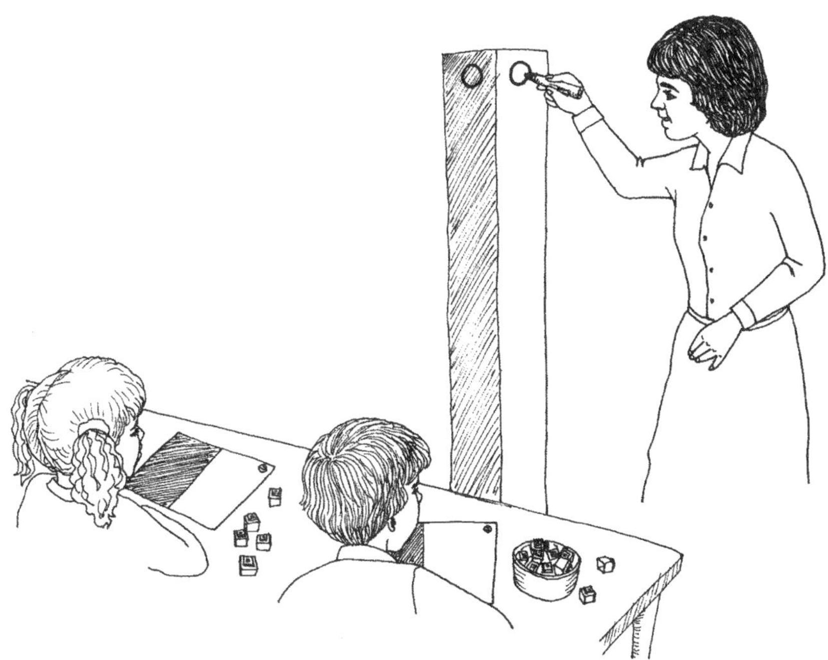

Zero gloms and zero.

* Based on *Mathematics Their Way,* "Counting Game: Using Symbols to Record the Concept," p. 299.

Plus one. How many?

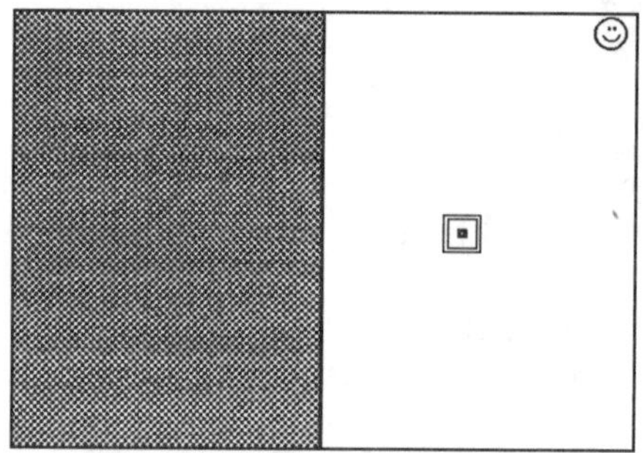

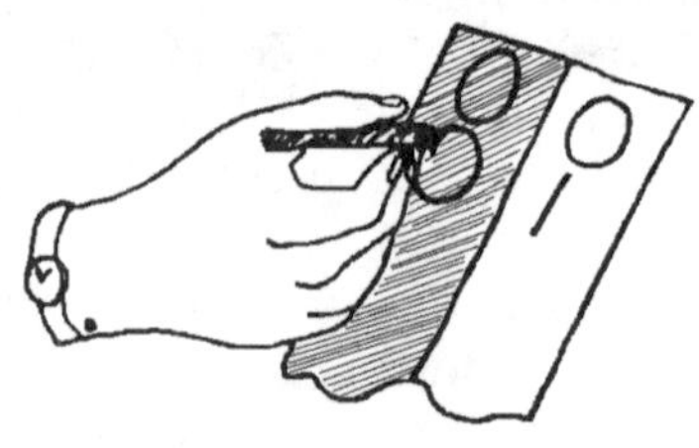

Zero gloms and one.

Plus one. How many?

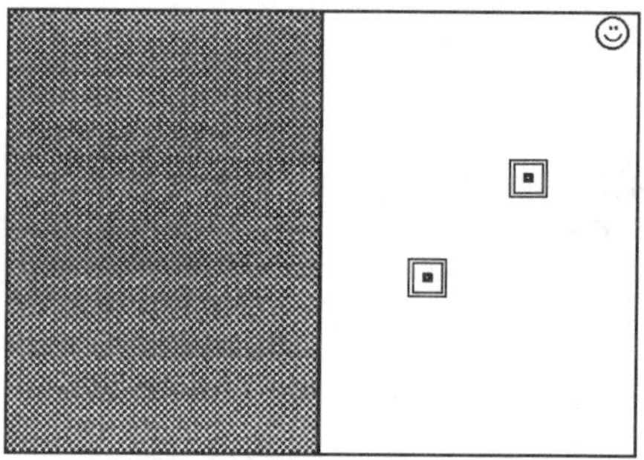

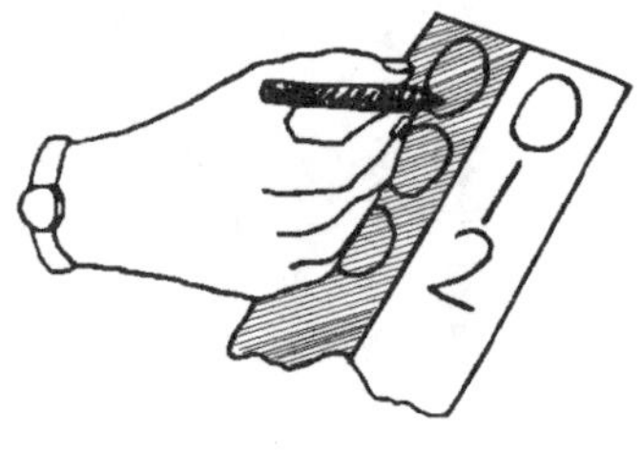

Zero gloms and two.

Continue until you reach four gloms and four. At that point, have the children look at the numbers in the white column and read them with you from top to bottom.

Who sees a pattern in the white column? What pattern do you see?

As the children identify each number sequence that repeats to form the pattern, draw a loop around it.

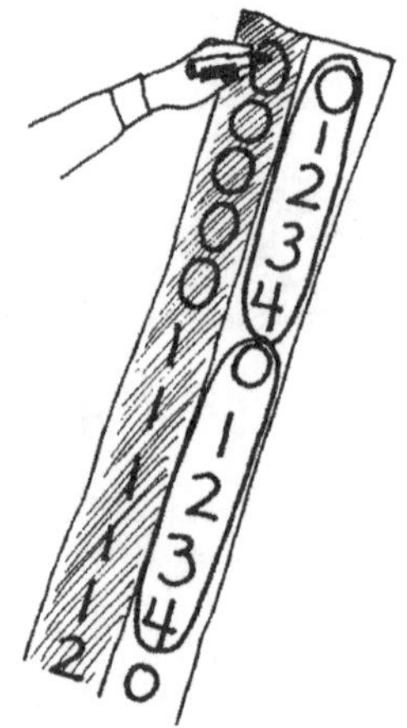

What pattern do you see in the green column? Let's read the numbers. Are some of the numbers the same?

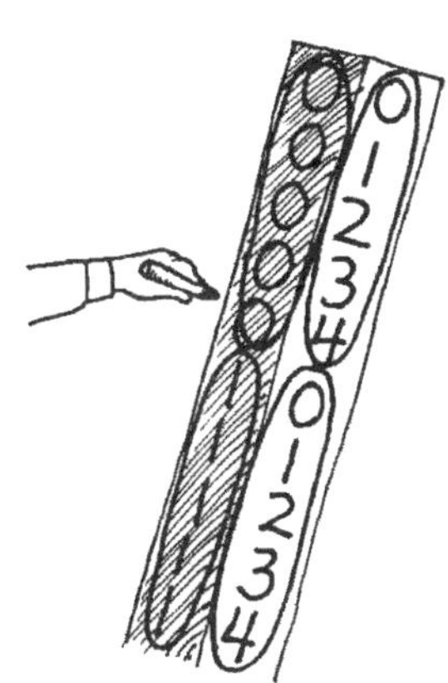

Zero, zero, zero, zero, zero; one, one, one, one, one; two, two, ...

Again, draw loops to highlight the number patterns as the children identify them.

After the patterns have been looped, have the children count the number of zeros, the number of ones, and so forth in the green column.

Minus-One Number Patterns

Give children a chance to discover the subtraction pattern that occurs as you remove cubes from the board. Have the children start with the maximum number of cubes on their place-value boards. (If they are working with "zibs," they would have three zibs and three.) As you lead them through the Minus-One grouping game, record on the long strip of paper the number of cubes they have on their boards at each step. For example:

Put three zibs and three on your board.

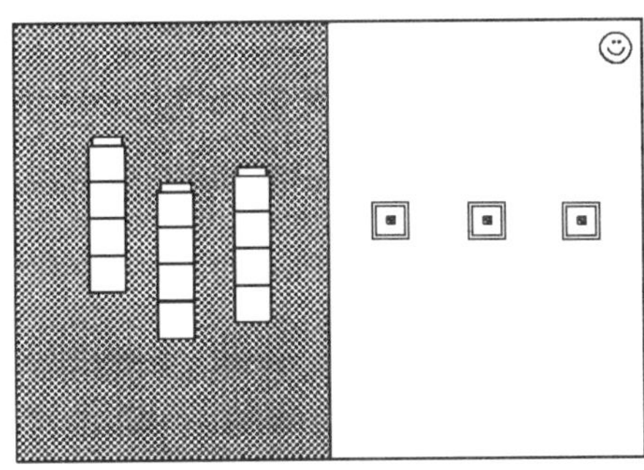

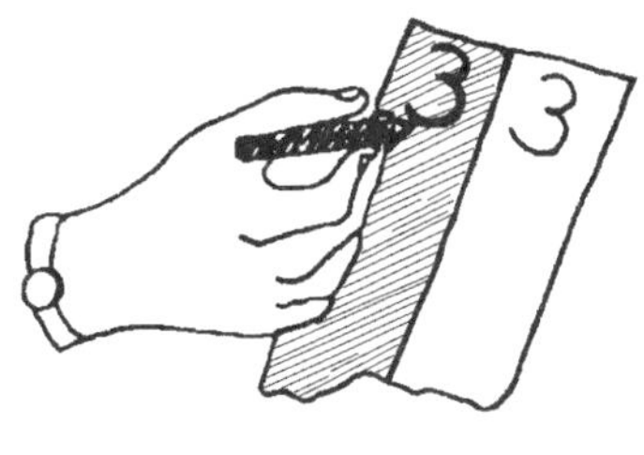

Minus one. How many?

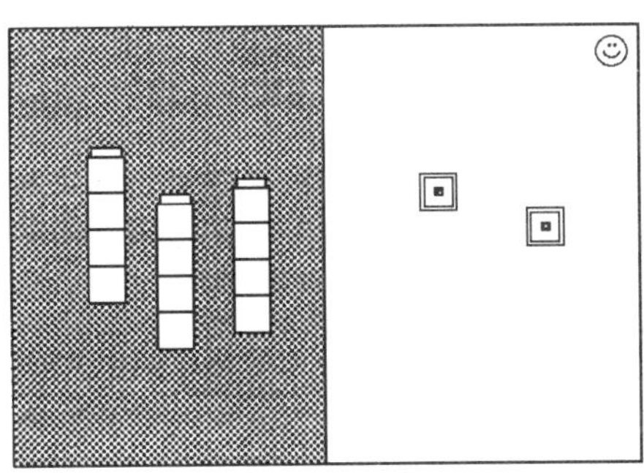

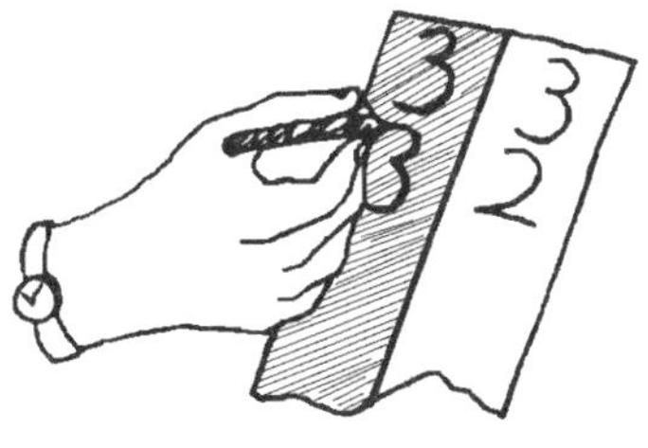

Three zibs and two.

Minus one. How many?

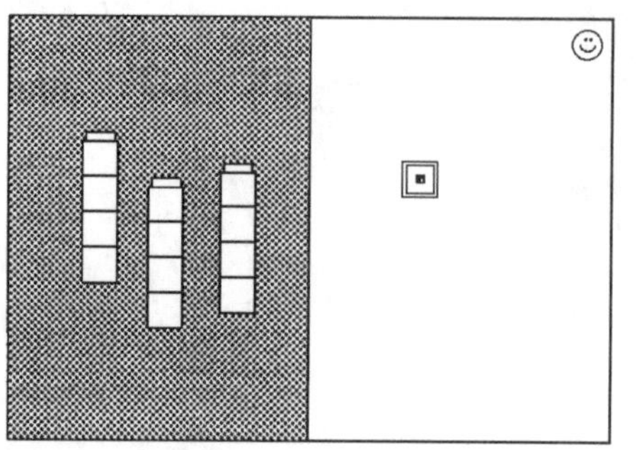

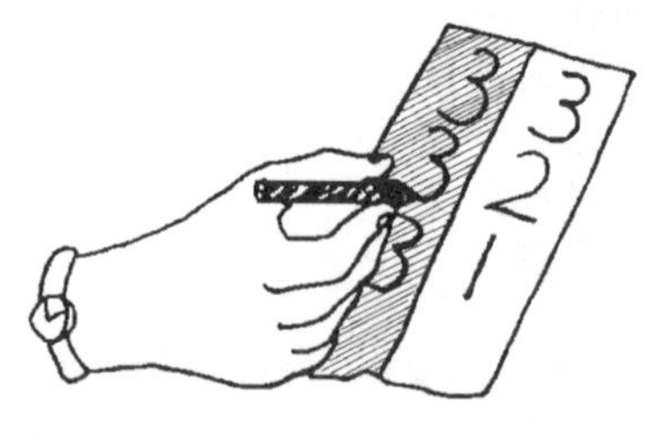

Three zibs and one.

Continue until you reach zero zibs and zero. Then have the children read the pattern in each column while you make the loops.

1–6 Recording the Plus-One and Minus-One Patterns .... Independent Activity

Materials: Connecting cubes (sorted by color) or beans and cups ▪ Place-value boards (1 per child) ▪ Place-Value Strips [BLM #113]

On succeeding days, have the children work independently to repeat the pattern-finding of activity 1–5 on their own. They can use the other groupings they have worked with and compare the various number patterns that occur. Have them first record the patterns on place-value strips and then loop the patterns.

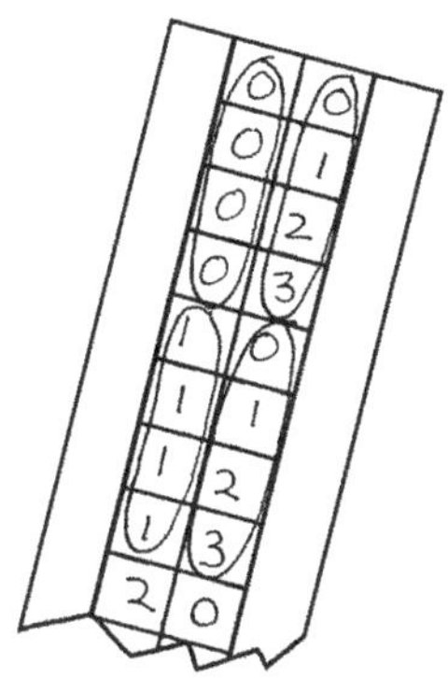

Extension: Plus-Two, Plus-Three, and Plus-Four Patterns

Some children will want to see what patterns emerge when they add more than one cube at each step. Allow them to explore on their own. Many will want to go beyond the two places on the place-value board. Remind them how to make the "big groups" by putting cube-trains into tubs on additional sheets of paper to the left of the place-value board. They can record these groups by writing in the margins of the place-value strips. (If working with bases other than base ten is a new experience for you, explore these patterns with your children. It's fun!) For example:

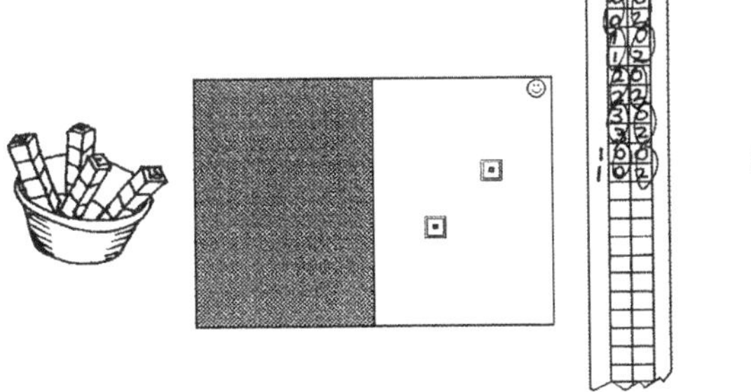

I am working with zibs.
I'm adding twos.

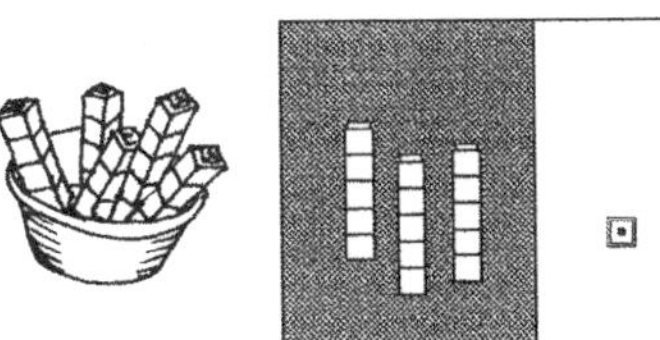

I am working with gloms.
I'm adding threes.

1–7 Introducing Number Patterns in a Matrix

Whole-Class Activity

Materials: Connecting cubes (sorted by color) or beans and cups ▪ Place-value boards (1 per child) ▪ Graph Paper [BLM #114]
Preparation: Prepare a variety of matrices for recording number patterns. Cut out a 4 × 4 matrix from the graph paper for children to use when playing with groups of four. Cut out a 5 × 5 matrix to use with groups of five, a 6 × 6 matrix to use with groups of six, and so forth.

As you play the grouping games with the children, use a matrix to record the number patterns that occur. For example:

Today we are going to play the zib game. As we play, I want you to tell me what to write.

How many are on our board now?

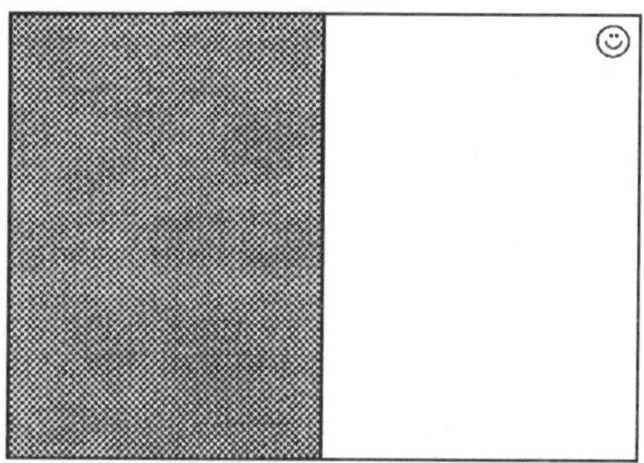

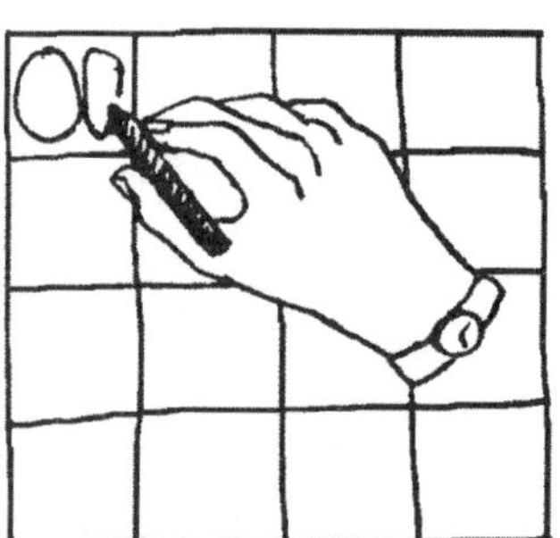

Zero zibs and zero.

Plus one. How many?

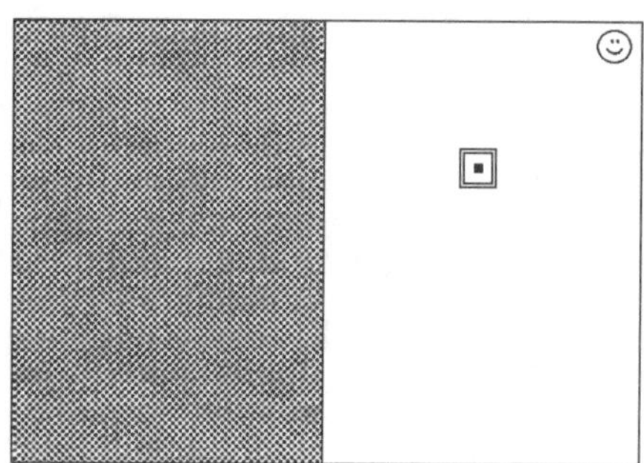

Zero zibs and one.

Plus one. How many?

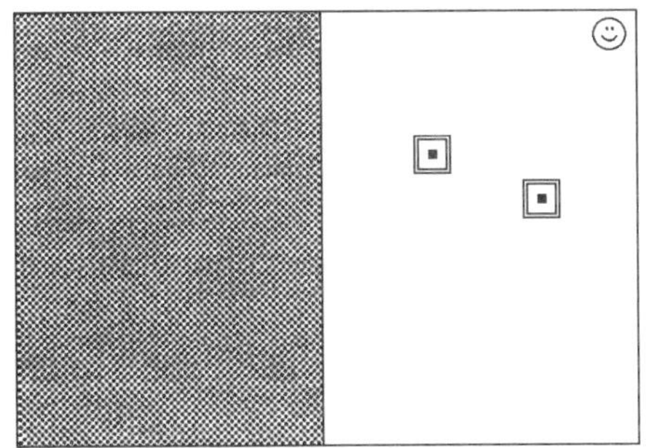

Zero zibs and two.

Continue until the matrix is completed. Then, have the children look for the patterns that occur in the matrix.

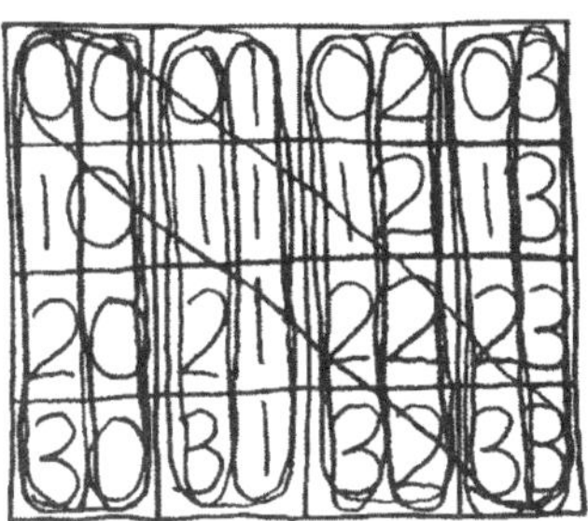

1–8 Recording the Patterns in a Matrix .... Independent Activity

Materials: Connecting cubes (sorted by color) or beans and cups ▪ Place-value boards (1 per child) ▪ Graph Paper [BLM #114]
Preparation: Prepare a variety of matrices for recording number patterns. Cut out a 4 × 4 matrix from the graph paper for children to use when playing with groups of four. Cut out a 5 × 5 matrix to use with groups of five, a 6 × 6 matrix to use with groups of six, and so forth.

On succeeding days, work with groups of five and six. Give the children the appropriate matrix and have them do their own recording independently.

Teacher-Directed Activities

About the Base-Ten Activities

The preceding introductory activities provide children with a foundation for understanding base-ten numeration. Once children understand the concept of regrouping and learn to find the patterns that evolve, they then explore these same ideas working with tens.

In the following activities the children work as a group or independently with a variety of number patterns, including those that evolve when adding by twos, fives, and tens. Because you will be working with large numbers, you will need lots of manipulatives. Thus, for these activities it is more practical for the children to work with beans and cups rather than with connecting cubes.

1–9 Introducing Grouping by Tens

Whole-Class Activity

Materials: Beans and cups or connecting cubes ▪ Place-value boards (1 per child)

Introduce the idea of "ten" as being a group.

Today we are going to work with a new group, a group of ten. Instead of making up a special word for ten, we are just going to call the group a ten. (Add the word "ten" to your chart.)

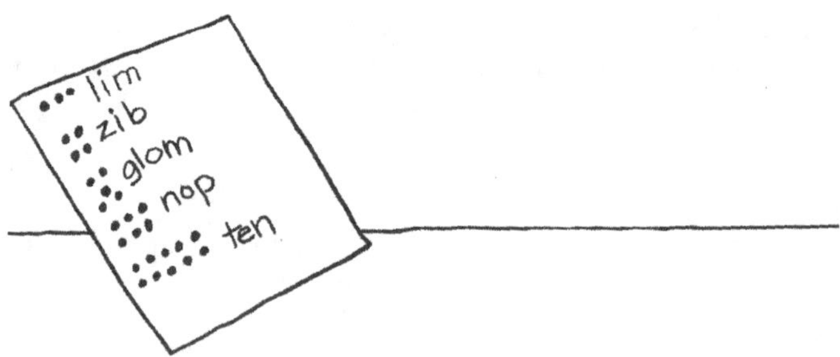

We are going to play the Plus-One game with ten. How many do we need to have on the white side of our place-value boards before we can put any on the green side?

Ten.

Yes, we will need ten. How many are on your board now?

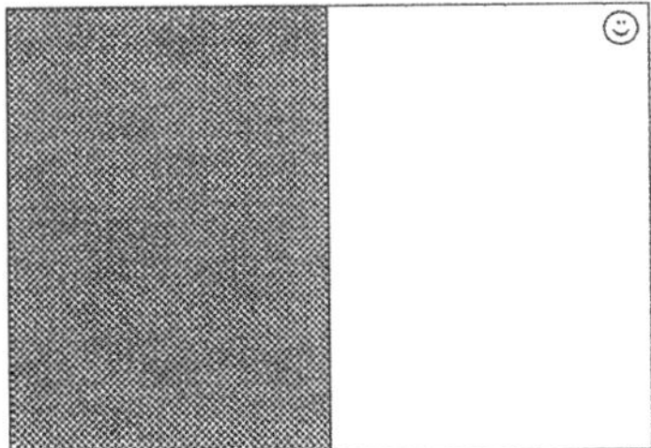

Zero tens and zero.

Plus one. How many?

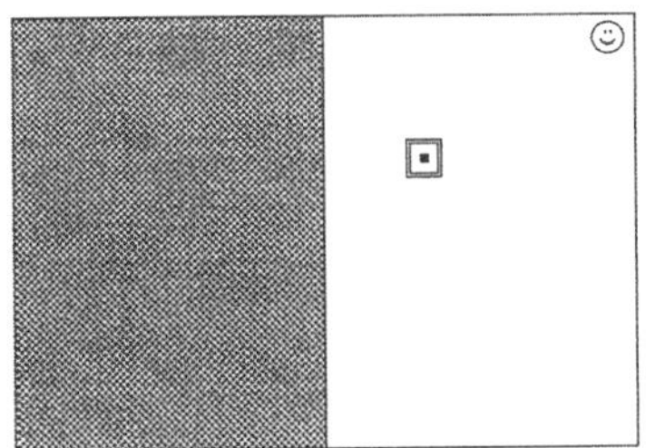

Zero tens and one.

Plus one.

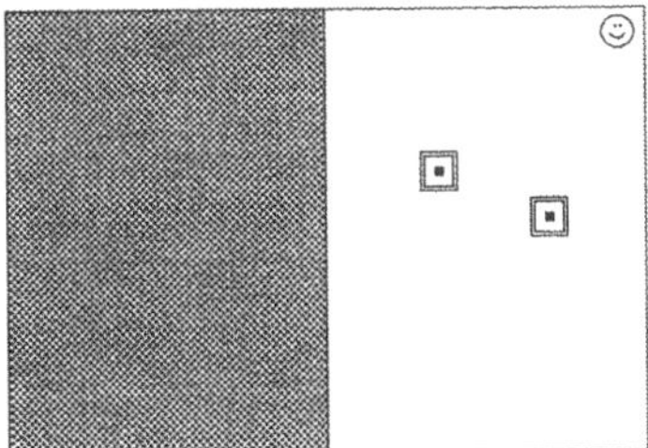

Zero tens and two.

Continue saying "Plus one" until you reach about three tens and four.

By this time it will be obvious why working with the smaller numbers was so useful for introducing regrouping. (It will take quite a while to reach even three tens and four with the children, and they will have regrouped only three times!)

Once you have introduced grouping with tens, have the children work independently on the following activity.

1–10 Writing Base-Ten Patterns on a Strip

Independent Activity

Materials: Place-Value Strips [BLM #113] ▪ Connecting cubes or beans and cups ▪ Place-value boards (1 per child) ▪ Clear tape ▪ 3" × 4" piece of tagboard for each child, rolled and taped ▪ Wooden clothespins

Have the children play the Plus-One game independently, using groups of ten. As they show each number on their board, have them write it on the place-value strip.

When the children finish a strip, they should use a pencil to draw loops around the number sequences that form the pattern on the strip. Then they should tape another strip to the bottom of the completed one so that they can continue recording numbers.

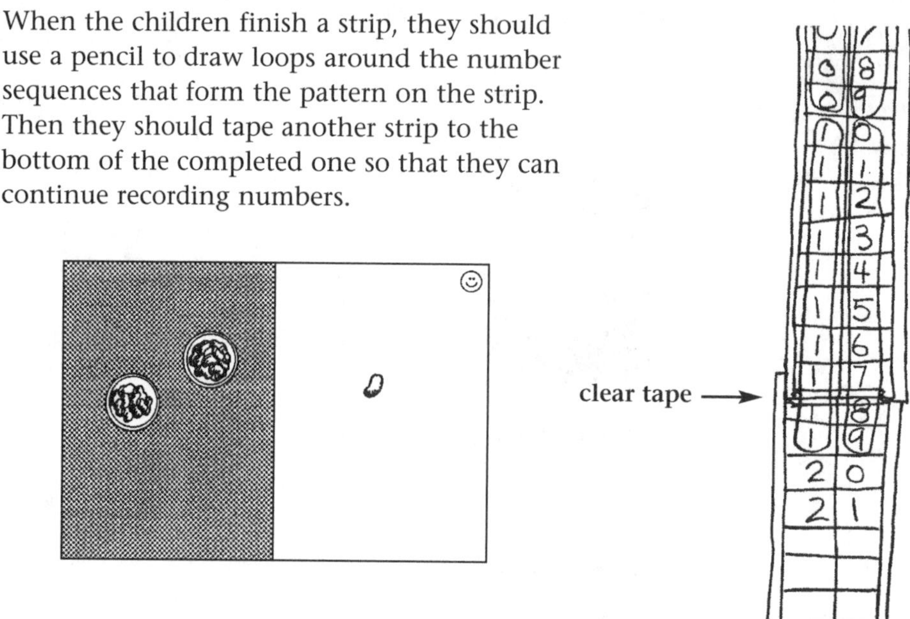

Show the children how to tape the top of their first strip to a tagboard roll. Then, as the strips are extended, they can be wrapped around the roll and held in place with a wooden clothespin. Write the child's name on the clothespin for identification.

The children should continue playing the Plus-One game, recording the numbers and extending strips until they reach the number 350. This will take several days. Many children get very involved in this task and like to see how far they can go. Encourage them to go as far as the time and materials allow.

When the children get to numbers beyond 99, they write the numeral representing the hundreds in the left margin of the strip. If they are using beans, they can also pour the ten cups of beans into a large envelope, writing "100" along with their name on the envelope.

At the end of a period, the children roll up their strips and put away the cubes or beans. Have them place cubes that have been snapped into tens into a container without breaking the tens apart. Cups of ten beans should be placed, intact, on a shelf. The next day, the children start by reading the last number on their strips, getting the necessary tens and ones manipulatives to build that number on their place-value boards, and continuing from that point.

This fresh start each day, requiring the children to use to manipulatives to interpret the last numbers on their strip, is one of the most important aspects of this task. For example:

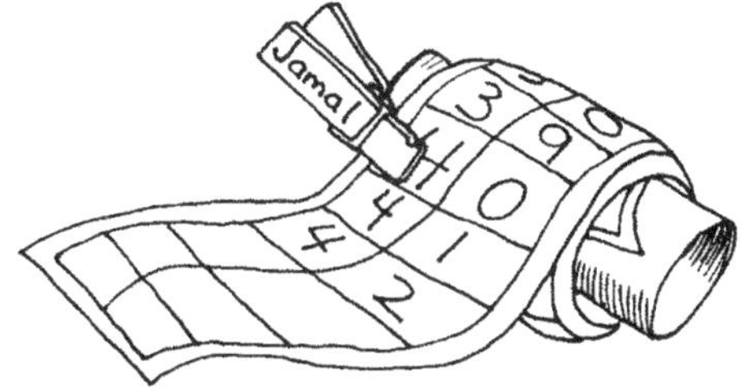

When the child starts work on the strip shown above, he or she first must read "42" and then show it with either cubes or beans and cups. This work reinforces the idea that 40, for example, represents four groups of ten and not just 40 individual objects.

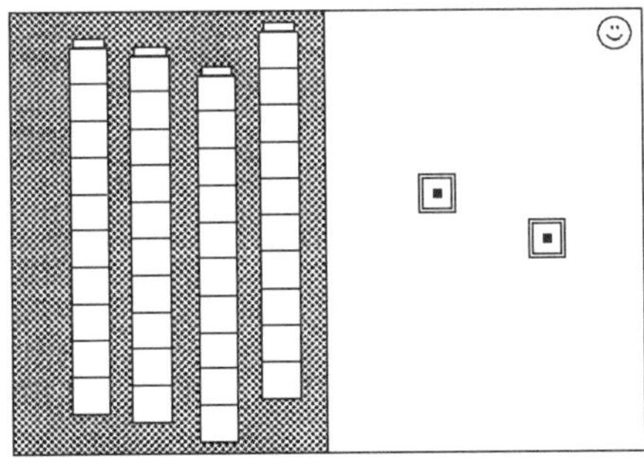

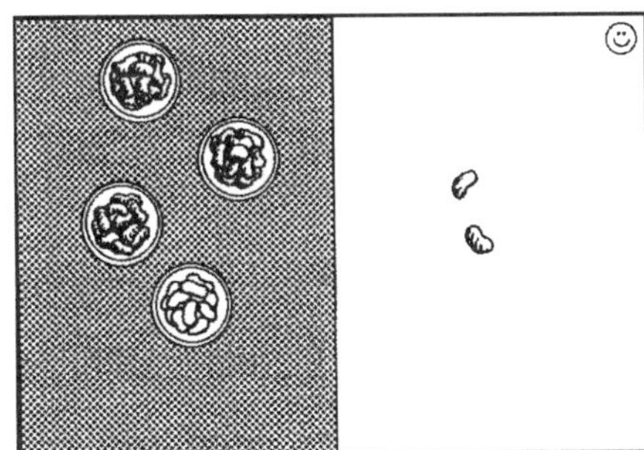

Continue to provide time over several days for the children to work. Have them continue to loop the patterns that evolve on their strips.

1–11 Creating a 00–99 Chart

Independent Activity

Materials: Connecting cubes or beans and cups • 10 × 10 Matrix [BLM #115]

As the children play the Plus-One game using base ten, have them each record the numbers on their own 10 × 10 matrix.

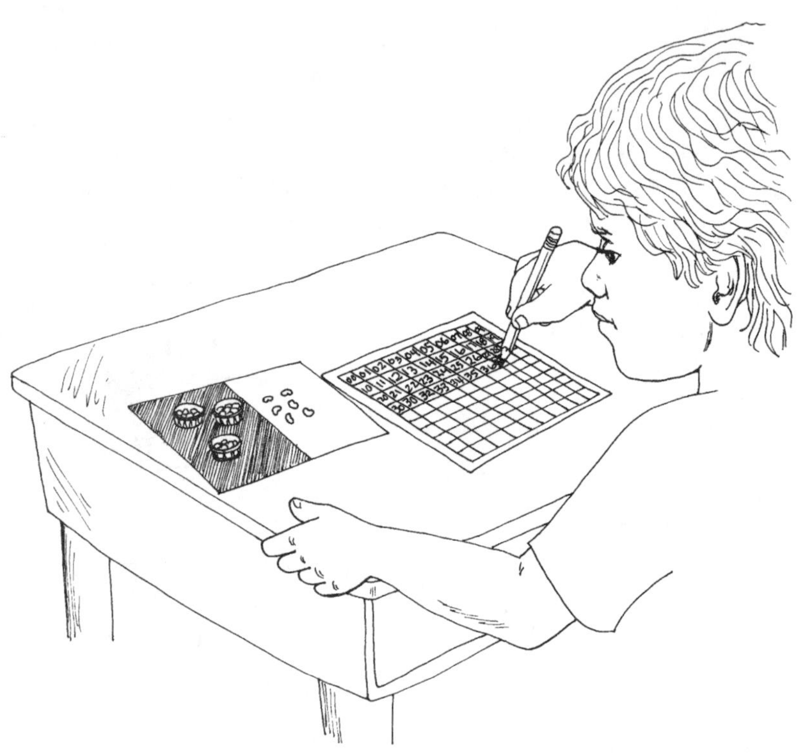

This will result in each child's having the meaningful experience of creating his or her own 00–99 chart.

1–12 Patterns on the 00–99 Chart

Whole-Class Activity

Materials: Large 00–99 chart for display, laminated or covered with acetate
Preparation: Prepare the 00–99 chart by marking off 100 two-inch squares on a large piece of tagboard.

Once the children have made their own 00–99 charts, hang up a poster-size 00–99 chart in the classroom. On occasion, discuss the number patterns that children point out and loop the patterns on the chart. Ask questions like these:

What pattern do you see in the columns? in the rows?

What patterns do you see in the diagonal lines?

Can you tell what number I am covering with my finger? How do you know?

For a related independent activity, see Looking for Patterns on the 00–99 Chart (activity 1–21).

1–13 Naming Patterns with Colors

Whole-Class Activity

Materials: Connecting cubes (sorted by color) or beans and cups ▪ Place-Value Strips [BLM #113] ▪ 00–99 Charts [BLM #116] ▪ Sheets of various colors of construction paper

Once the children have explored the base-ten patterns that emerge when adding by ones on a strip or matrix, they should begin adding groups of various sizes, such as by twos, threes, and so forth. They will then discover different patterns that appear.

Later, as the children work with patterns in a variety of settings, they will discover that these same patterns appear and reappear. To help them focus on and describe these recurring patterns, have the class choose a color to label each pattern they identify. For example, when the class talks about the 2, 4, 6, 8, ... pattern, they might choose to label it *red* (as described in the classroom scene that begins on page 10). When they find the 5, 10, 15, 20, ... pattern, they might name it the *blue* pattern. To introduce children to this idea, have them explore two or three different patterns and label them. For example, you might have the class add by twos, recording the pattern on place-value strips. Once the pattern has been identified as the 2, 4, 6, 8, ... pattern, it can be labeled *red*. The numbers that form that pattern can be written on strips of red construction paper and hung in the room for reference. The pattern can also be colored on a 00–99 chart which can be hung next to the strip.

On another day, children might explore the pattern that emerges when they count the fingers on each hand of everyone in the classroom. This five-pattern can be written on strips of *blue* construction paper for display and the pattern colored in on a 00–99 chart.

Later, as new patterns are discovered, new colors may be chosen to represent them. Each time the children find a pattern, they should check to see if it is the same as one they have already found. If it is a new pattern, they should assign it a color.

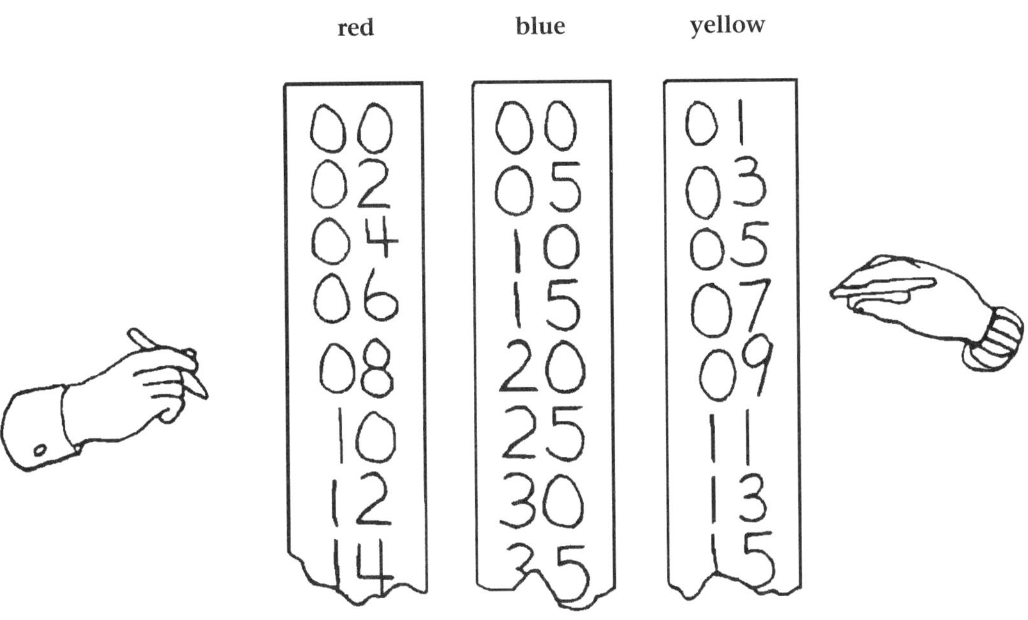

About the Number Patterns in Growing Patterns

It is important for children to see that number patterns occur in many different settings in the real world. One way for them to experience this is through their work with growing patterns. Activities that introduce growing patterns are included in Book One of this series.

If working with growing patterns is new to children, they will need some time to work independently. They will need to copy and extend patterns on task cards and create their own patterns before they are asked to label the parts of the patterns with numbers and to look for number patterns as in Book One, activity 2–17. If the children have had previous experiences with growing patterns, then they can analyze the patterns and move on quickly to working with number patterns.

1–14 Analyzing Growing Patterns

Materials: Connecting cubes, wooden cubes, or Color Tiles

When they first experience growing patterns, children extend patterns that you model. Then they do some independent exploration of patterns that grow. When you reintroduce growing patterns in the context of place-value activities the children will reexamine the patterns and discover a variety of number patterns. Encourage children to analyze and describe these patterns in their own ways before having them look for number patterns. For example:

I am going to begin a pattern and I want you to tell me how the pattern grows.

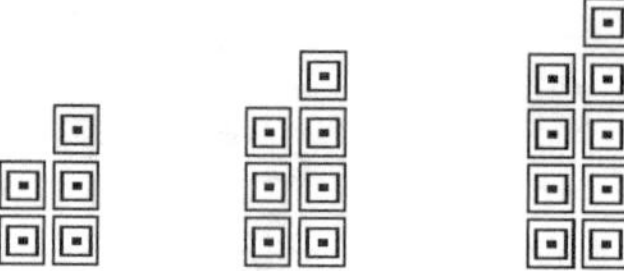

> **I see two and two, with one on top.**
>
> **Then I see two and two and two with one on top.**
>
> **Then I see four twos with one on top.**

Who looked at it another way?

> **I saw two going down and the three going down.**
>
> **The next one has three going down and then a four.**
>
> **Then four and five.**

Who has a different idea?

> **I saw two and two with one more.**
>
> **Then I saw three and three with one more.**
>
> **Then four and four with one more.**

Build the next step—How did you know what to build? What comes next? and next? How do you know?

The following are examples of patterns you might work with:

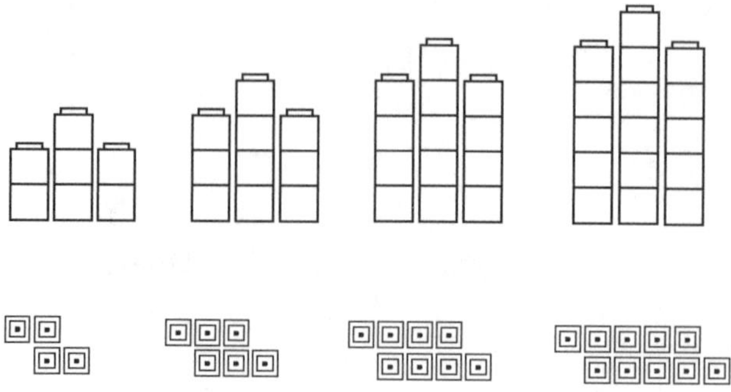

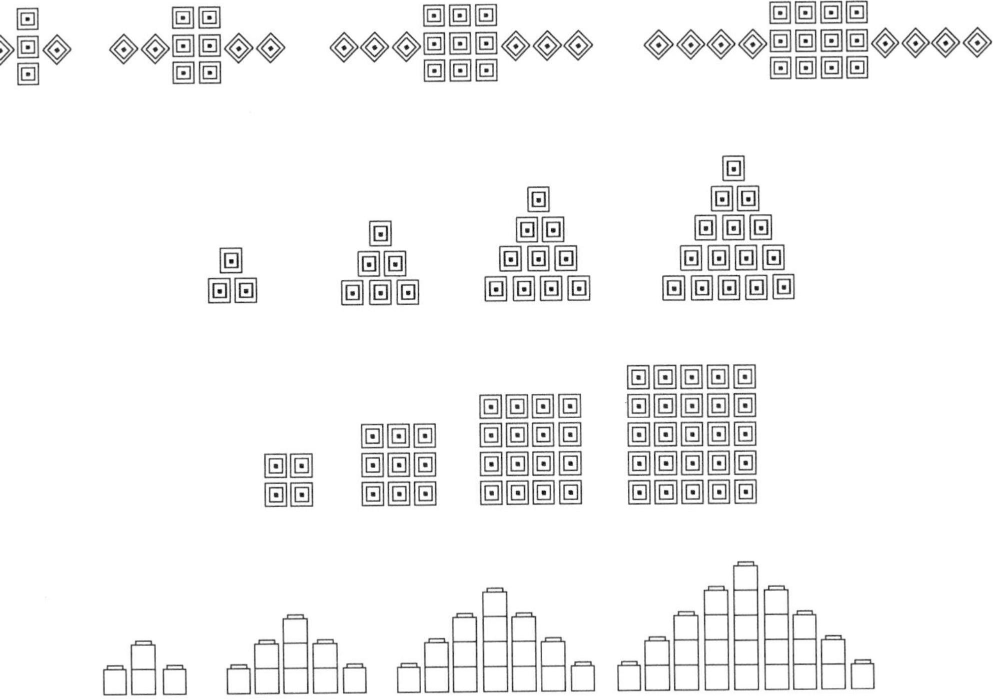

NOTE: A single design may grow in different ways as long as its growth pattern is consistent. For example, here are two different ways of extending patterns made from a single design.

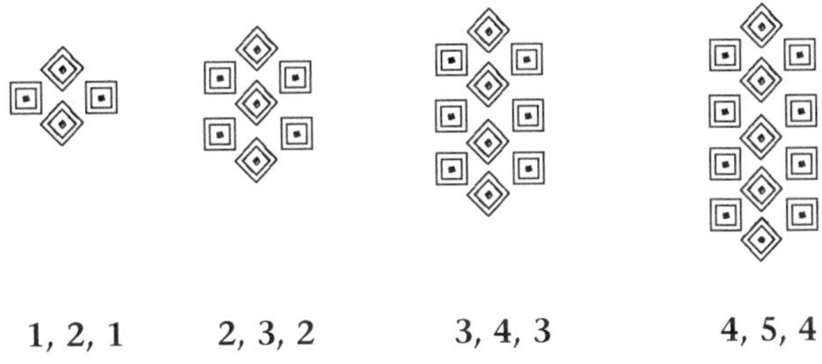

or

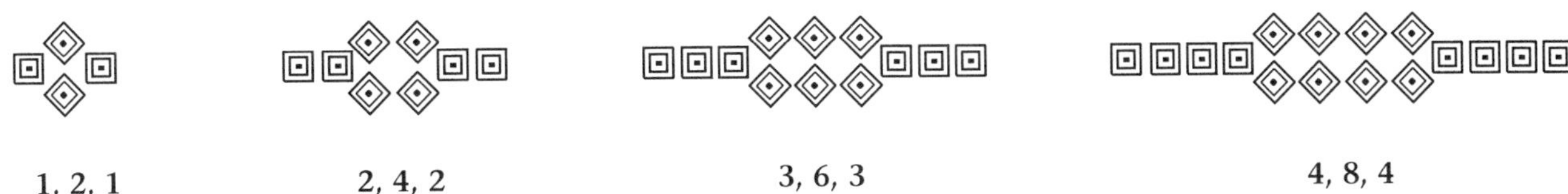

1–15 Finding Number Patterns in Growing Patterns

Whole-Class Activity

Materials: Connecting cubes or Color Tiles • Two large 00–99 charts for display • 00–99 Charts [BLM #116] • Crayons

In the preceding activities, the children analyze growing patterns as they determine how to make the designs grow in a consistent way. Now they will reexamine their patterns in new ways to discover a variety of number patterns.

For example, in the following designs the children can look for number patterns occurring in the "arms," middle, and total number of cubes used in each step.

What pattern can you find when you look at the "arms" of these designs?

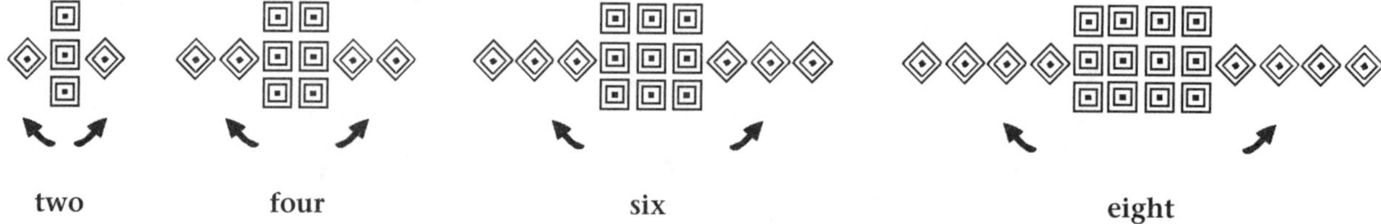

As the children report the numbers, circle them on a large 00–99 chart.

00	01	02	03	04	05	06	07	08	09
10	11	12	13	14	15	16	17	18	19
20	21	22	23	24	25	26	27	28	29
30	31	32	33	34	35	36	37	38	39
40	41	42	43	44	45	46	47	48	49
50	51	52	53	54	55	56	57	58	59
60	61	62	63	64	65	66	67	68	69
[illegible]	[illegible]	[illegible]	[illegible]	74	[illegible]	76	77	78	79

After children have reported all the numbers they see, have them predict the numbers that would make up the arms of the design if they continued to extend it. Continue to circle those numbers on the 00–99 chart. Doing this demonstrates the power of pattern: the fact that you can make predictions beyond the physical evidence you have at present.

00	01	02	03	04	05	06	07	08	09
10	11	12	13	14	15	16	17	18	19
20	21	22	23	24	25	26	27	28	29
30	31	32	33	34	35	36	37	38	39
40	41	42	43	44	45	46	47	48	49
50	51	52	53	54	55	56	57	58	59
60	61	62	63	64	65	66	67	68	69
70	71	72	73	74	75	76	77	78	79
80	81	82	83	84	85	86	87	88	89
90	91	92	93	94	95	96	97	98	99

Now ask children to identify the growing number pattern found in the middle of the designs and discuss how the numbers would continue to grow if children extended the pattern. Circle the numbers they report on the other large 00–99 chart.

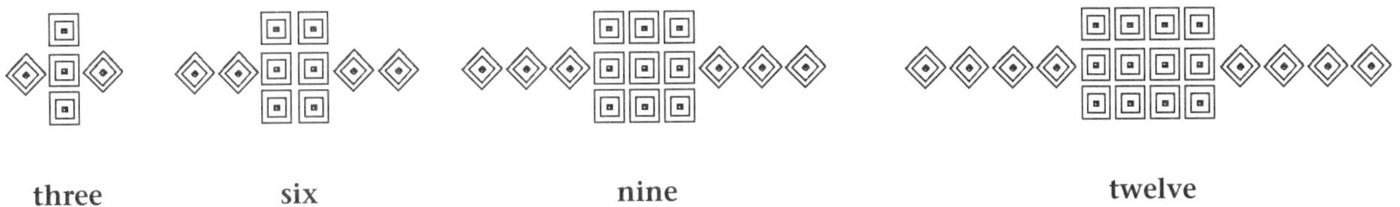

three **six** **nine** **twelve**

00	01	02	03	04	05	06	07	08	09
10	11	12	13	14	15	16	17	18	19
20	21	22	23	24	25	26	27	28	29
30	31	32	33	34	35	36	37	38	39
40	41	42	43	44	45	46	47	48	49
50	51	52	53	54	55	56	57	58	59
60	61	62	63	64	65	66	67	68	69
70	71	72	73	74	75	76	77	78	79
80	81	82	83	84	85	86	87	88	89
90	91	92	93	94	95	96	97	98	99

Repeat the discussion for the growing pattern formed by the total number of cubes in each step. Record the numbers children report.

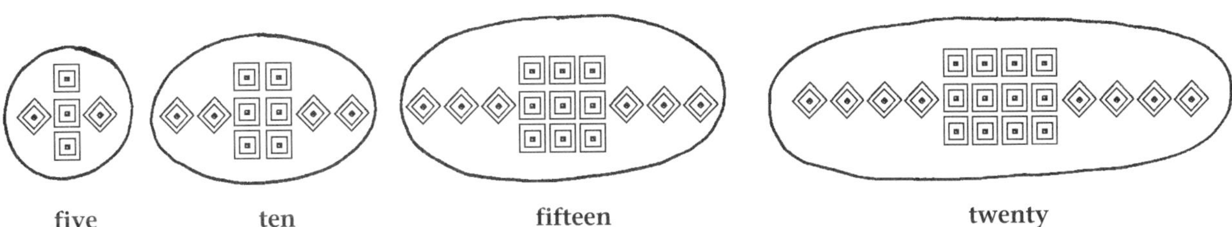

five **ten** **fifteen** **twenty**

About Color Labeling of the Number Patterns

As the children work, they will discover a variety of growing patterns. Later they may rediscover those same patterns in other situations. Some of these may be patterns they found while working with activities using the place-value strips (with the Plus-Two, Plus-Three, and Plus-Four games). If they find a pattern that matches one already on display on a color strip, they use a matching crayon to color that pattern on a small 00–99 chart.

Imagine, for example, that you are working with children on the following growing pattern.

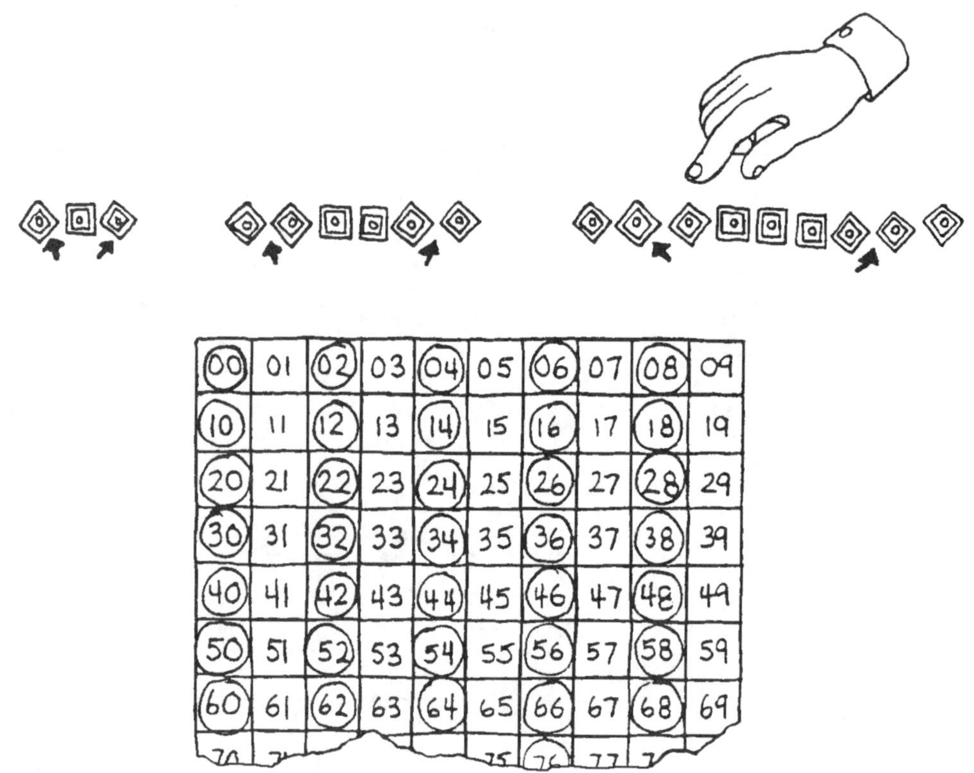

The children discover that the pattern for the "arms" is a 2, 4, 6, 8, ... pattern.

Circle the pattern on a 00–99 chart as the children predict what the pattern would look like if it were extended.

Does this remind you of any of our color patterns?

> **Yes, it's like the red pattern. The red pattern goes 2, 4, 6, 8, ... too.**

On succeeding days, the children will make more designs, extend them, and continue to look for patterns. If they discover a number pattern that is the same as one of those already posted, they will identify the pattern as that particular one. If they discover a new pattern, they will choose a new color and new strips will be made for display in the room.

1–16 Introducing Pattern Searches .... Whole-Class Activity

Materials: 12" × 18" sheet of construction paper • 00–99 charts (one for each pattern you explore) [BLM #116] • Crayons

The children explore, identify, and then sort various patterns according to the colors the class has previously decided to use. (See Naming Patterns with Colors, activity 1–13.)

In this ongoing activity, as children discover patterns in new situations, they will begin to see the connections between patterns and will notice when the same patterns emerge, even when the situations are different. For example, they may discover that the pattern for counting eyes is the same pattern as for counting wheels on bicycles. The pattern for counting numbers of legs on ladybugs comes up again when children see how many green pattern-block triangles fit on a sequence of yellow pattern blocks. Experiencing the same pattern over and over again and becoming familiar with it will also prepare children for work with multiplication.

Introduce the search for patterns by choosing a particular pattern and using a sheet of construction paper to record it in various ways. For example:

What pattern do we find when we count eyes?

Call on one child at a time to come to the front of the room while the class determines the number of eyes for each number of children. Model how to keep track of the pattern by making a table.

One child has how many eyes?

Two eyes.

Two children have ———?

Four eyes.

Three children have ———?

Six eyes.

Four children have ———?

Eight eyes.

Now that we have a few steps of our pattern down, can anyone predict how many eyes five children have? six children? ten children? How do you know?

What does the eye pattern look like on the 00–99 chart? (Tape a small 00–99 chart onto the construction paper.)

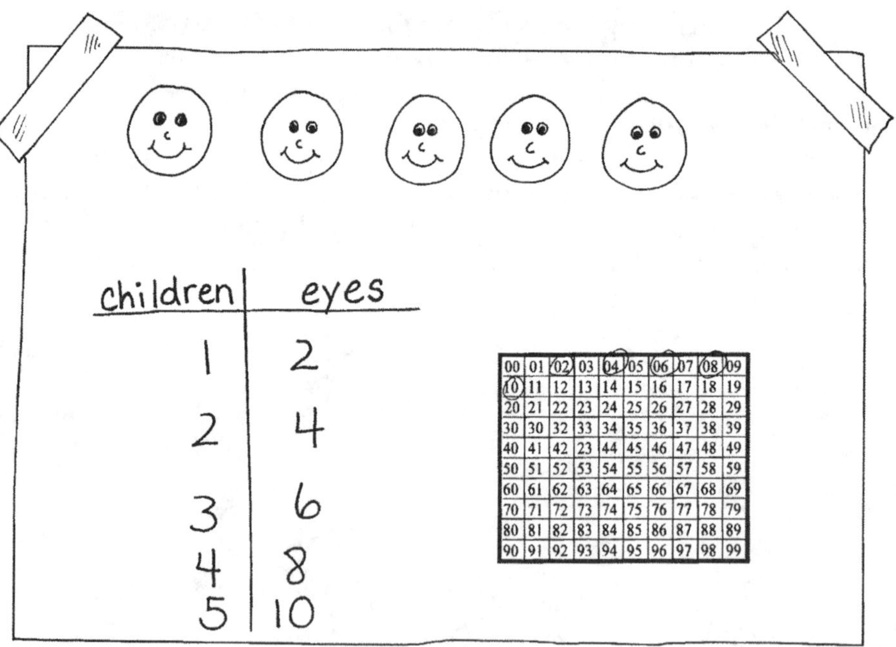

Is this like any other pattern we have worked with before? What color did we decide to give to this pattern?

That looks like the red pattern.

So we can trace the numbers of the pattern with a red crayon, and we can color the pattern on the 00–99 chart as well.

Let's try something else and see what pattern we get this time. Let's make snowmen. How many large snowballs do we need to make one snowman? two snowmen? three snowmen?

We can record the pattern this way:

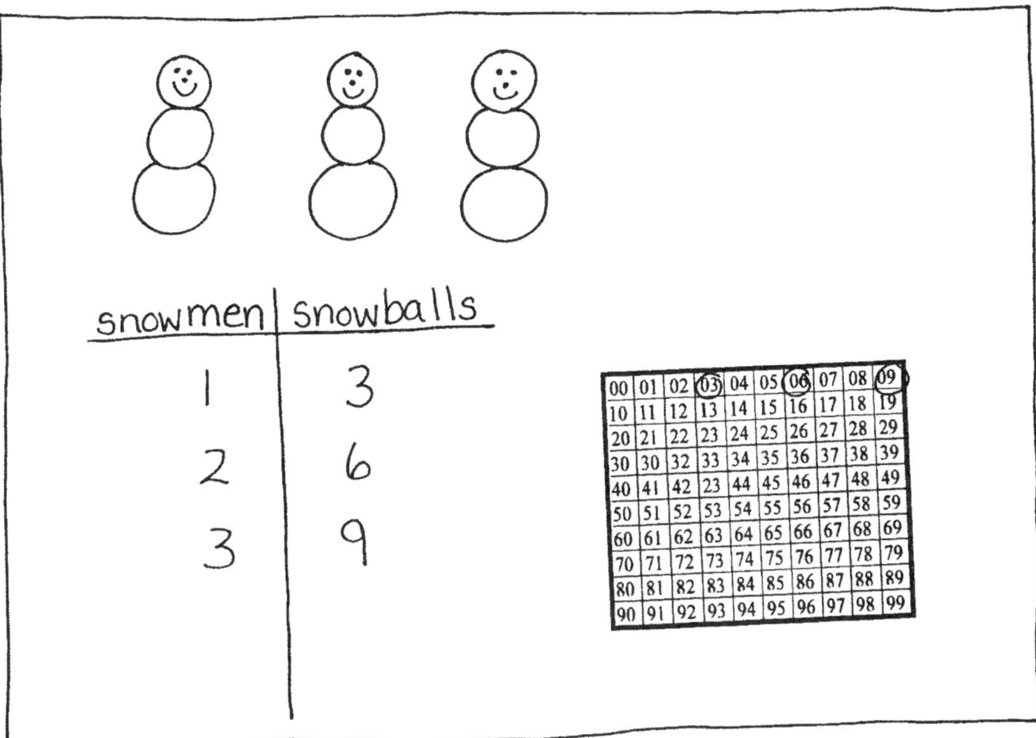

Explore various number patterns in this way throughout the year.

For a related independent activity, see Searching-for-Patterns Station (activity 1–23).

Independent Activities

About Number-Pattern Stations

The following activities can be presented as a set of independent stations providing children with a variety of activities with which to explore base-ten number patterns.

1–17 Recording Various Number Patterns on Strips

Independent Activity

Materials: Connecting cubes or beans and cups • Place-value boards (1 per child) • Place-Value Strips [BLM #113]

Have the children add by twos, threes, fives, tens, and so forth. Have them record each count on a place-value strip, looping the pattern that evolves.

+2

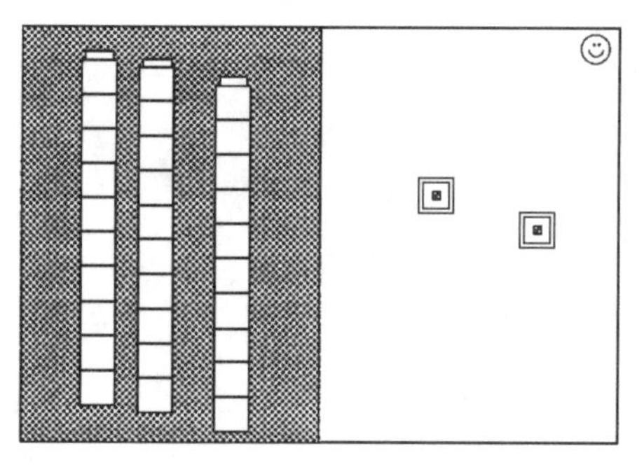

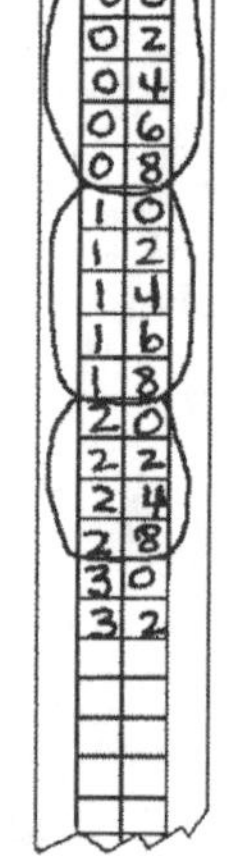

+6

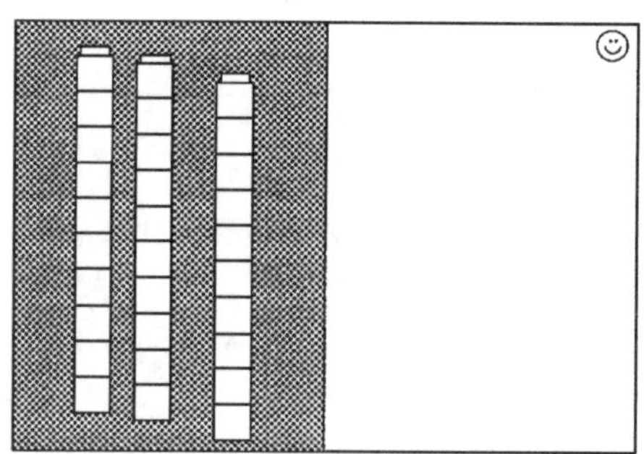

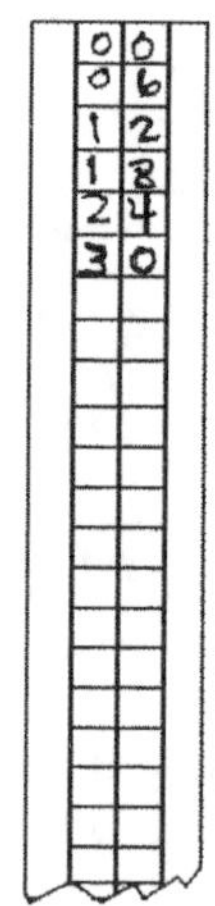

Children should then refer to the color patterns displayed in the room and decide whether or not their pattern matches any of the ones posted. If they discover a new pattern, they should report it to the class so that another color can be chosen to represent it.

1–18 Grab and Add .... Independent Activity

Materials: Connecting cubes or beans and cups ▪ Place-value boards (1 per child) ▪ Place-Value Strips [BLM #113] ▪ 0–6 Number cubes (1 per child, optional)

Have each child grab a handful of cubes or beans, arrange them into tens and ones, and place them in the appropriate spaces on the place-value board. The child then chooses a number (or rolls a number cube) and repeatedly adds that number of counters to the place-value board, recording the resulting number pattern on a place-value strip. This activity will help children discover that whatever number they start with can be used to begin a number pattern.

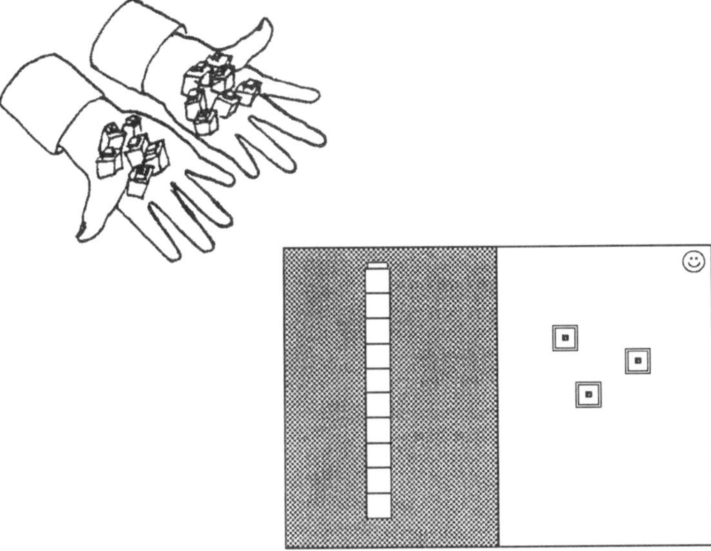

I had one ten and three ones.

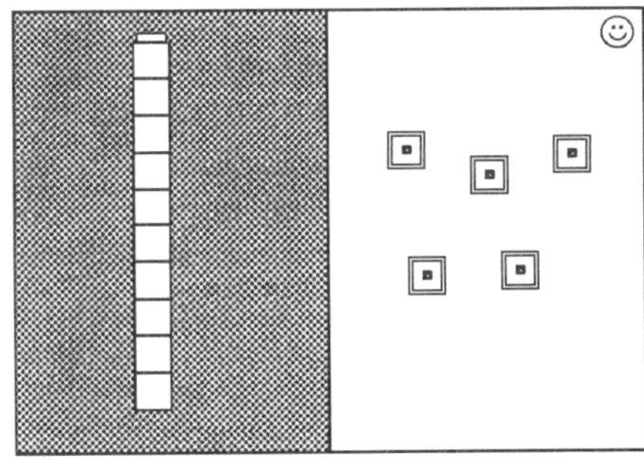

I will add two each time.

1–19 Number Patterns in Growing Patterns

Independent Activity

Materials: Connecting cubes, wooden cubes, or Color Tiles ▪ Shelf paper or butcher paper (36" long) ▪ 1-inch-square pieces of paper ▪ Glue ▪ 00–99 Charts [BLM #113]

The children create growing patterns and record them by gluing paper squares onto a long strip of paper. They find a number pattern, label it, and then circle the pattern on the 00–99 chart. Then they glue the 00–99 chart onto the long strip of paper.

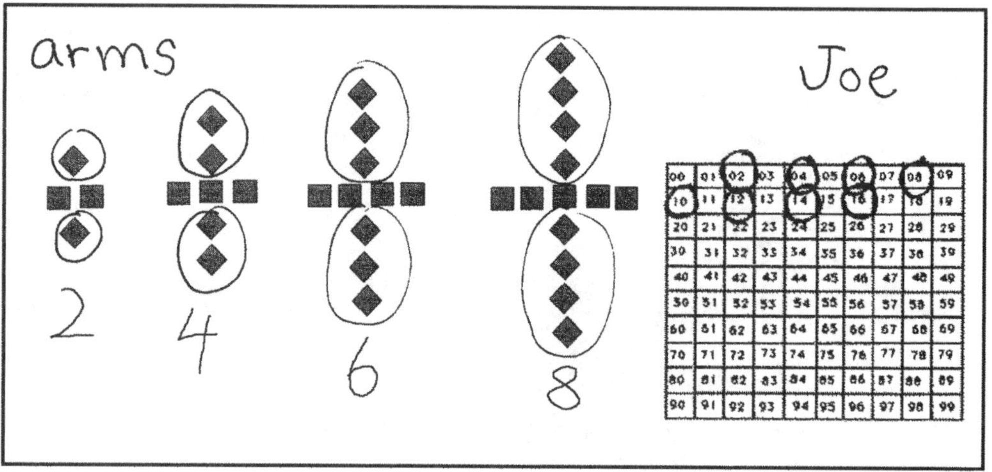

1–20 Margie's Grid Pictures*

Independent Activity

Materials: Connecting cubes ▪ Grid Picture Task Cards [BLMs #118–125] ▪ 00–99 Chart (Duplicate BLM #116 on tagboard and laminate.) ▪ 10 × 10 matrix (1 per child) [BLM #115]
Extension: 10 × 10 matrices with just a few numbers filled in. (Duplicate BLM #115 and fill in a few of the numerals as shown below.)

Children practice finding numbers on the 00–99 chart. Then they discover some number patterns in the pictures they create by placing connecting cubes directly on the chart.

Children place cubes of different colors on the 00–99 chart on those numbers indicated by the task card. When the listed numbers are covered in the appropriate colors, a picture appears on the chart.

**My friend Margie Gonzales invented these grid pictures when she was teaching first grade, so I have always thought of these as "Margie's grid pictures."*

Red

4 13 14 15 22
23 24 25 26 31
32 33 34 35 36
37 40 41 42 43
44 45 46 47 48
50 52 54 56 58

Black

64 74 84 94 82
92 93

Grid Picture Task Card H

It's an umbrella!

Extension: Using Grids with Missing Numbers

Children can be challenged by having to place the cubes on a 10 × 10 matrix that has some of the 00–99 numerals already filled in.

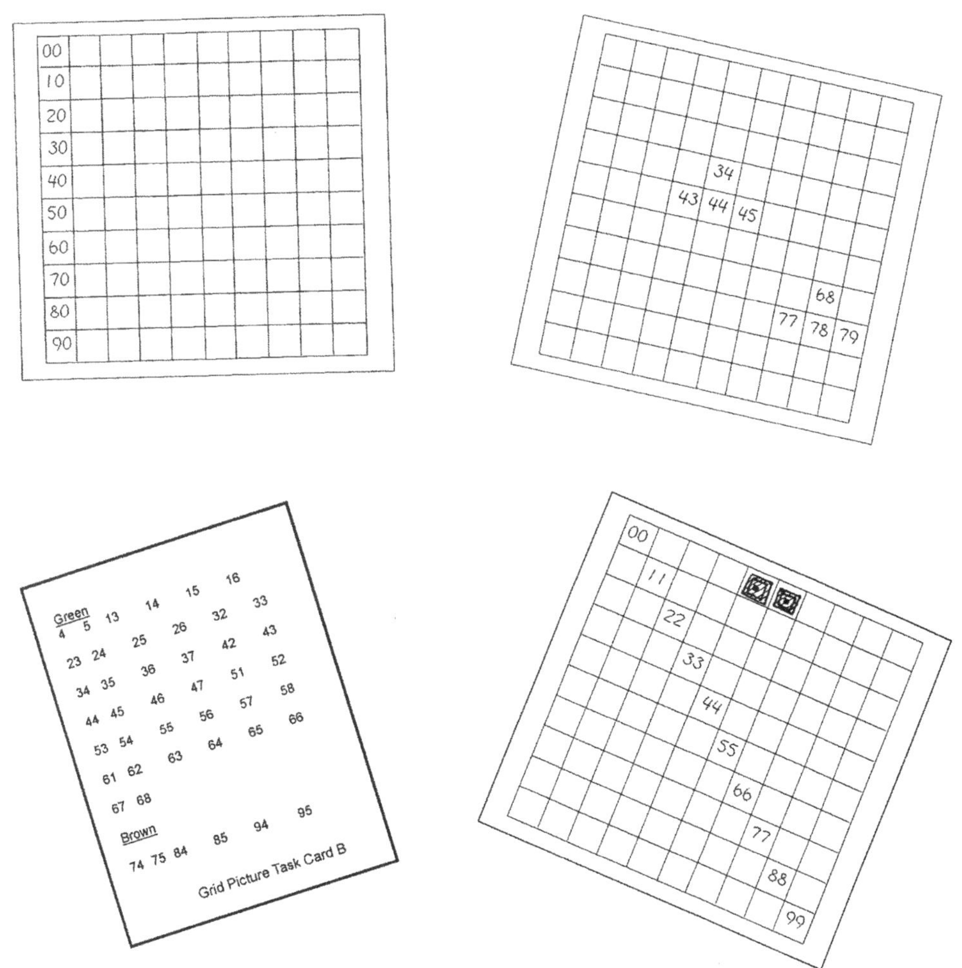

I need to figure out where 13 is.

1–21 Looking for Patterns on the 00–99 Chart

Materials: *Level 1:* Connecting cubes ▪ 00–99 Charts (Duplicate BLM #116 on tagboard and laminate.)
Level 2: Same as for Level 1, but add 00–99 charts duplicated on regular paper and crayons.
Extension: Prepare grids other than the 00–99 charts. (See pictured examples.)

This activity gives children the opportunity both to encounter some of the number patterns they have experienced previously on the 00–99 chart and to discover new patterns.

Level 1: Making Number Patterns

The child names a number from two to ten and places a connecting cube on that numeral on the 00–99 chart. The child who names "four," for example, places a cube on the numeral 4. Then, starting with the next space to the right, the child counts to four over and over again, placing a cube on every fourth square. For example:

00	01	02	03		05	06	07		09
10	11		13	14	15		17	18	19
	21	22	23	24	25	26	27	28	29
30	31	32	33	34	35	36	37	38	39
40	41	42	43	44	45	46	47	48	49
50	51	52	53	54	55	56	57	58	59
60	61	62	63	64	65	66	67	68	69
				74		76	77	78	79

I'm adding by fours.

On another day the child names a different number, places a cube on the numeral for that number, and then counts to that number over and over again, placing cubes on the chart to determine a different pattern. For example:

00	01	02		04	05		07	08	
10	11		13	14		16	17	18	19
20	21	22	23	24	25	26	27	28	29
30	31	32	33	34	35	36	37	38	39
40	41	42	43	44	45	46	47	48	49
50	51	52	53	54	55	56	57	58	59
60	61	62	63	64	65	66	67	68	69
				74		76	77	78	79

I'm adding by threes.

Level 2: Recording Patterns by Color

The child colors in the pattern on the 00–99 chart to reflect the colors of a pattern already displayed in the classroom. (See Naming Patterns with Colors, activity 1–13.)

Extension: Patterns on Other Grids

Have the children repeat the activity using smaller or larger grids. For example:

00	01	02	03	▣
05	06	07	▣	09
10	11	▣	13	14
15	▣	17	18	19
▣	21	22	23	24

I'm adding by fours.

00	01	02	03	▣	05	06	07
▣	09	10	11	▣	13	14	15
▣	17	18	19	▣	21	22	23
24	25	26	27	28	29	30	31
32	33	34	35	36	37	38	39
40	41	42	43	44	45	46	47

I'm adding by fours, too.

1–22 The 00–99 Chart Puzzles

Independent Activity

Materials: 00–99 Puzzles (See preparation below.) ▪ Crayons ▪ 00–99 Chart for reference (optional)

Preparation: To make the 00–99 puzzles, start by duplicating BLM #115 for a supply of 10 × 10 matrices. Cut apart each grid in various ways. (See examples below.) Write one number on each puzzle and leave the other squares blank. Laminate each puzzle so that after a child is finished using it, it can be wiped off and reused.

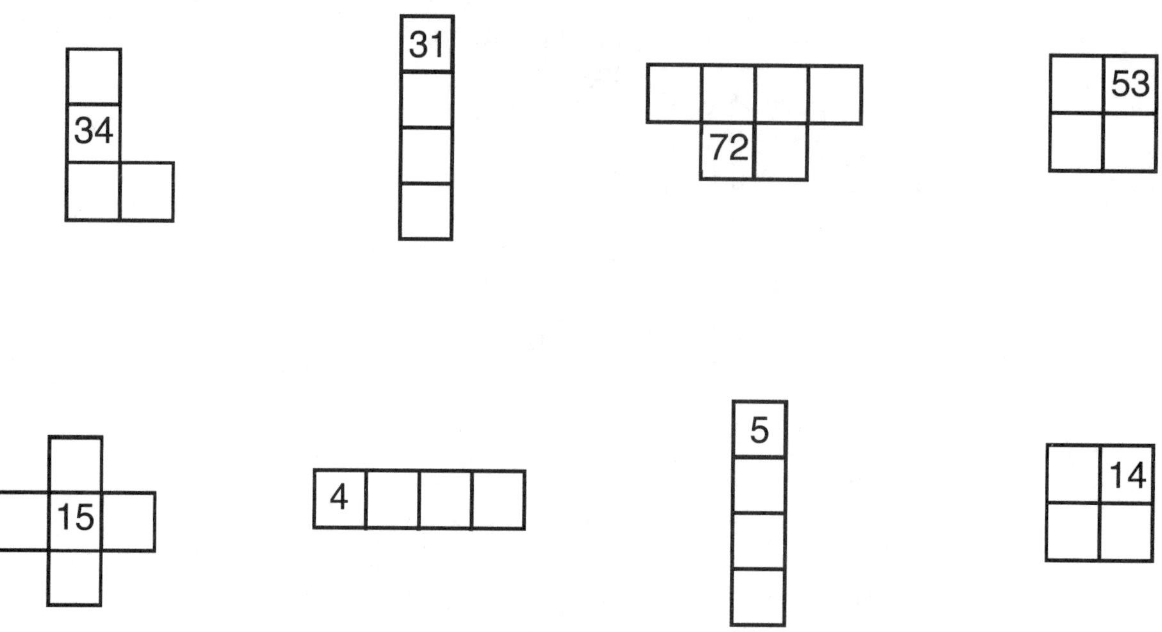

The children choose a puzzle and fill in the missing numbers.

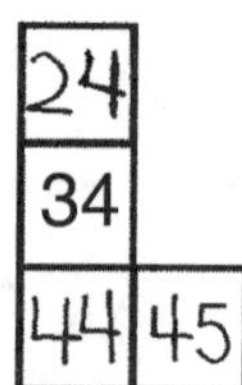

1–23 Searching-for-Patterns Station .... Independent Activity

Materials: 12" × 18" Sheets of construction paper ▪ Hole punch ▪ Metal rings (1 per child) ▪ Crayons ▪ 00–99 Charts [BLM #116] ▪ List of questions for children to explore (See examples below.)

In the teacher-directed activity, Introducing Pattern Searches (activity 1–16), you modeled different ways for the children to record and label patterns they find around them. Now the children can work independently to discover, record, label, and sort a variety of patterns, using "T" tables, 00–99 charts, and their own drawings. Have some questions listed for the children to explore. Also encourage them to explore their own ideas. For example:

I found the pattern for eyes.

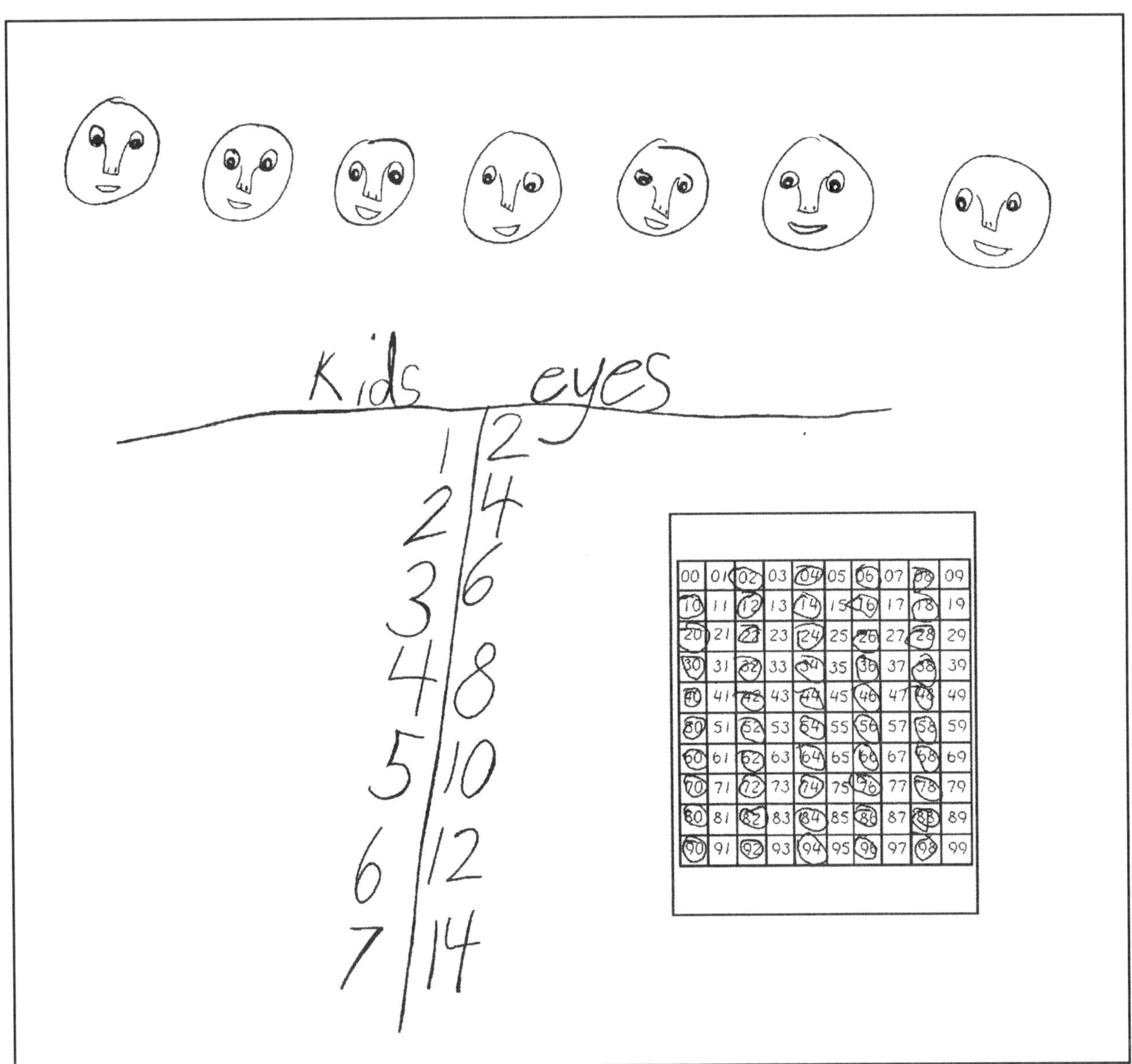

I found the pattern for whiskers. It is the green pattern.

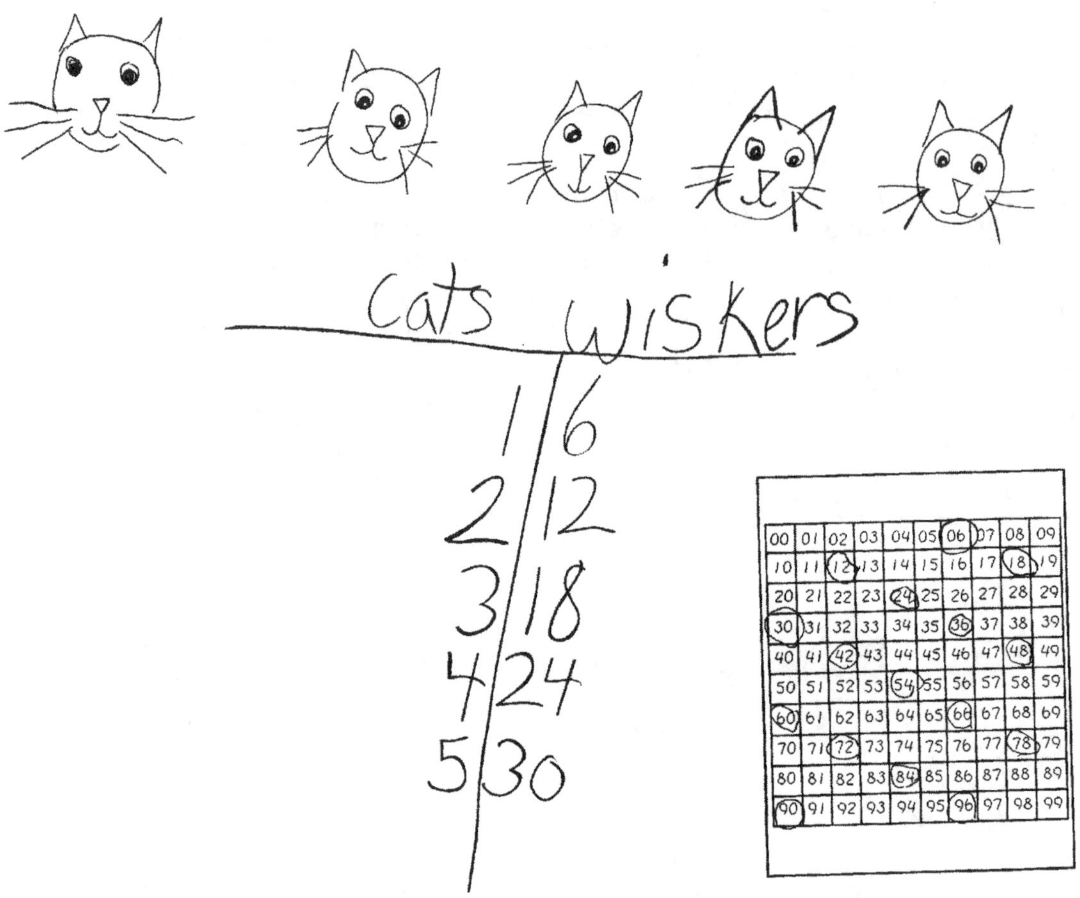

As children find and sort their patterns over time, they can create their own pattern books. (Punch a hole in one corner of each page and insert a metal ring to hold the pages together.)

The following are pattern questions you can list for children to explore:

- What pattern do we find when we count the number of sides of triangles? of squares? of diamonds? of hexagons?
- What pattern do we find when we count the number of wheels on our toy cars?
- What pattern do we find when we count the number of legs on chickens? on horses?
- What patterns can we think of that would go with our blue pattern? our red pattern?
- What pattern do we get when we count bicycle wheels? tricycle wheels?

SECTION B

Developing a Sense of Quantities to 100 and Beyond

The activities in this section help children understand the value and importance of organizing numbers into tens and ones. They also help children to think of numbers in terms of tens and ones and to develop a sense of quantity for numbers to 100.

Goals for Children's Learning* *(Section B)*

Goals

When counting numbers to 100 in a variety of settings, the children will:

- Make reasonable estimates
- Develop an understanding of conservation of large numbers
- Use their knowledge of tens and ones to determine *how many* without counting by ones
- Record quantities as numbers of tens and ones
- Take numbers apart in a variety of ways
- Compare two quantities to see which is more and which is less
- Compare two quantities to determine how many more one number is than another

When working with the activity extensions, the children will:

- Organize quantities into hundreds, tens, and ones
- Record quantities as numbers of hundreds, tens, and ones

Analyzing and Assessing Children's Needs

When we want to assess our children's competence in handling place-value concepts in order to make decisions about what they need to learn, it is not enough to know if they can name the digits in the tens and ones places in a number. We need to know if they can interpret these numbers correctly. We need to know whether or not they can organize large quantities into tens and ones to determine amounts. Thinking of numbers in terms of groups of tens and "leftovers" is an important foundation for adding and subtracting two-digit numbers. The following questions will help you recognize the stages that children are moving through as they develop an understanding of numbers to 100.

* Adapted from *How Do We Know They're Learning? Assessing Math Concepts.*

Questions to Guide Your Observations*

Questions

Taking Numbers Apart

- Can the children organize objects into groups of tens and leftovers in a variety of ways to model a particular number?
- When working with a particular number, after they have decided that there are a certain number of tens, do they predict the number of leftovers, or do they count to find out? For example, if they have a pile of 35 cubes and they make one 10, do they know that there will be 25 leftovers?

Conservation of Number

- Do the children know that a given quantity does not change when it is regrouped, or do they need to count to make sure?

Estimating

- Are the children reluctant or at ease about making estimates when working at the independent place-value stations?
- Can they use the information they get as they work and change their estimates as they get more information? Are their new estimates closer to the actual amount?
- Are they comfortable with (and honest about) their estimates, or do they try to change them to match their answers?

Using Tens and Ones to Determine How Many

- Do the children organize their counters into tens and ones? What methods do they use? Are they able to record the quantities appropriately?
- Do they determine an amount by counting by ones? by counting by tens? by looking at the tens and leftovers?

Comparing Quantities

- Can the children compare quantities to determine which is more and which is less?
- Can they figure out how many more one number is than another? What size numbers are they confident in using? Are they able to figure out how many more or less when the difference is ten or less? when the difference is greater than ten?
- What methods of comparing do they use? Do they use counters to compare, or do they compare numbers? Do they count up from one number to another? Do they subtract one number from another?

* Adapted from *How Do We Know They're Learning? Assessing Math Concepts.*

Working with Symbols

- Are the children able to record the same number in a variety of ways? For example, can they write 43 as 4 tens and 3, as 43, and as 40 + 3?
- Are they able to interpret various ways of recording numbers accurately?

Meeting the Range of Needs

The independent activities in this section (for use at place-value stations) are naturally expandable to meet a range of needs. As you watch the children working at the stations, you will see them approaching identical tasks in many different ways. Some children will not see the value of organizing counters into tens and leftovers and will count them all by ones. You will see some children counting all the counters by ones even if they have organized them into tens and leftovers. You will see others counting by tens and ones. Still others will be counting and changing their estimates as they gather new information. The tasks can also be easily adapted to provide appropriate challenges as children become ready for them. Children can be asked to compare quantities, to find out "how many more" one quantity is than another, and to work with numbers larger than one hundred. (See page 66 for more information.) If any children need a challenge, you can suggest that they make a comparison. Thus, all these tasks have continuous value when experienced repeatedly over a long period of time.

A Classroom Scene

The children have been working for several weeks with the place-value stations. When they first started, they all had the same eight activities to work with, including Lots of Lines (1–32), Paper Shapes (1–33), Containers (1–36), and Measuring Things in the Room (1–38). At this point, the teacher has been able to identify and accommodate the different levels at which the children are working.

Independent Work: *Place-Value Stations*

Today Mr. Bautista moves around the room to observe the children at work. He stops first at the Paper Shapes station (activity 1–33), where Colleen and Juan are each filling a paper shape with Color Tiles. Colleen has organized the red and yellow tiles she is using into groups of ten of each color. Juan is filling his paper shape with blue tiles and is marking each ten with a red tile.

Mr. Bautista asks Juan how many cubes he has used so far. Juan glances at his paper shape and says, "Three tens and five—that's 35."

"What about yours?" the teacher asks Colleen.

Colleen starts counting by tens to find out. "Ten, twenty, thirty, forty, forty-one, forty-two, forty-three—forty-three," she responds.

Eddy is also working with paper shapes. Mr. Bautista brought out some of the shapes for the numbers between 10 and 20 because Eddy was having difficulty dealing with these larger numbers. "How many tiles do you have on your shape so far, Eddy?" Mr. Bautista asks.

Eddy counts each of the tiles. "Sixteen, so far," he says.

Dominique has two paper shapes. She has filled each one and has written the quantities on her How Many More? worksheet. Now she is trying to determine how many more one paper shape holds than the other.

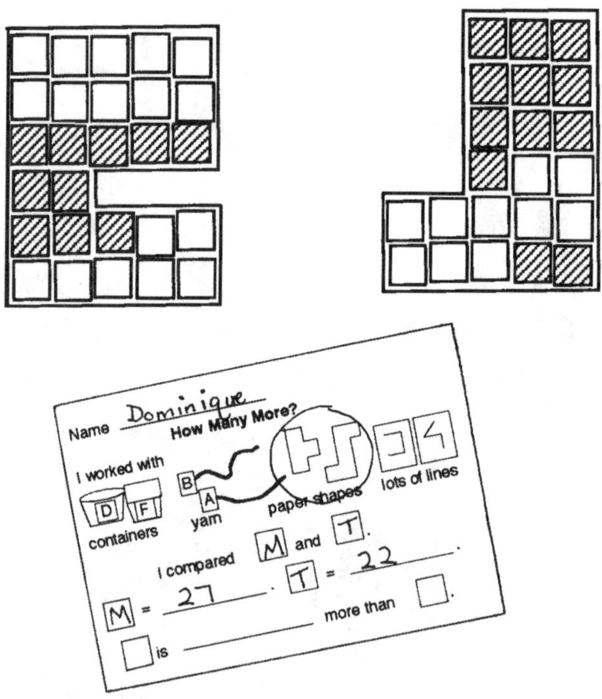

When Mr. Bautista demonstrated how to use this worksheet a few days ago, he said that whoever wanted to work with comparing could pick up one of the worksheets. Dominique and Breyonna are the only ones who have been interested so far. Mr. Bautista plans to have everyone work at that level eventually, but he wants to provide the challenge right now for any children who are ready for it.

A top priority for Mr. Bautista is to find the right level of work for each of the children. He doesn't want some to feel that math is too easy, or others to feel that it is too hard. The children have come to accept the fact that their classmates can work on different tasks at the same time. The teacher has worked hard to create an environment in which the children just accept this. One of Mr. Bautista's techniques is to remain matter-of-fact when the children move on to higher levels. He is pleased to see the children working hard at whatever level they are working, and he conveys that to them.

Mr. Bautista used to think it was motivating to children to see others moving ahead and being praised for it. Now, he has come to believe that children are motivated to work hard when tasks are at the level that is appropriate for them. Mr. Bautista knows some children are inclined to give up if a task is too difficult for them, so he plans carefully to make sure that everyone can do the tasks they are assigned. But he also wants everyone to have to work hard. If he finds children working at a level that appears to be too easy for them, he will pose a challenge to redirect them. In fact, most children want to work at the appropriate level. They don't necessarily see the various tasks as being at different levels. Instead, they see them as different questions they are exploring.

The teacher moves over to the group working with Containers (activity 1–36). He has supplied lots of different items, including apricot pits and small pompoms, which the children are using to fill a variety of small containers. The numbers the children work with at this station are generally larger than those they work with at most of the other stations. While the teacher is watching the other children put counters into little cups to make tens, he notices that Adrian is doing something different. "What are you doing, Adrian?" he asks.

"I got tired of putting these in cups, so now, every time I get to ten, I just put a counter down and that stands for a ten," Adrian tells him. Adrian has expressed the beginning of the idea that one object can stand for many. Mr. Bautista is pleased to see that Adrian has figured this out for himself. Judging from past experiences, he expects to see several other children learning this process from Adrian.

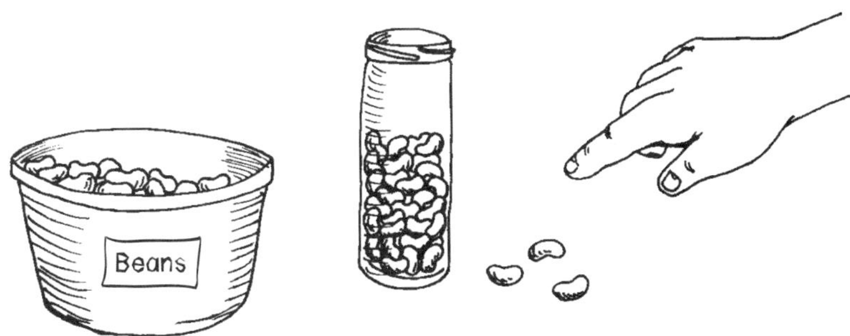

Yukio and Sam come over to Mr. Bautista all excited. "We made a train all the way from this end of the rug to the other end. It was 432 cubes long!"

Mr. Bautista is pleased to see their excitement. They were quite challenged by working with the large number of cubes. "How did you figure out how long it was?" he asks them.

"After we built it, we broke all these into tens," says Yukio, holding out some of the trains of ten she made. "Then we could count them all by tens."

Sam joins in the conversation. "And I can count them by fives, too," he says proudly.

Sam then proceeds to try to count each of his trains of ten by fives: "Ten, fifteen, twenty, twenty-five, thirty, thirty-five."

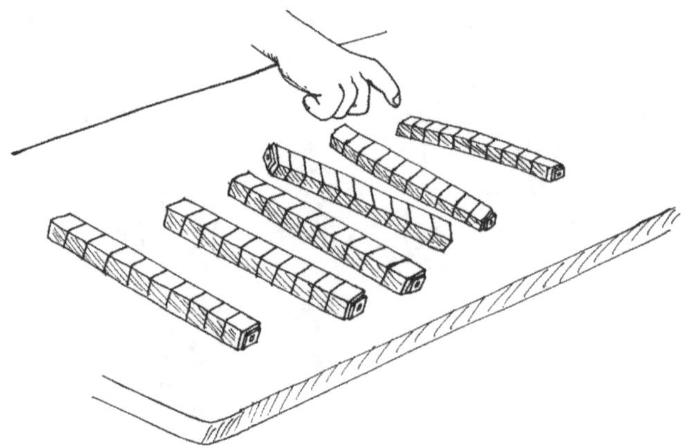

Even after working with children for many years, Mr. Bautista is surprised by the way Sam counted. He knows that these ideas are complex and that children often get confused or misled, even when using models. He knows that many second graders, because they have not yet developed an understanding of conservation of large numbers, believe that the numbers change when they are counted in different ways. Mr. Bautista realizes that just telling Sam he is wrong won't help him understand this idea. On the other hand, he doesn't want to let this go without commenting, either. So he says to Sam, "Counting by fives means we count five at a time. How many fives do you see in each train?"

Sam knows Mr. Bautista is trying to correct him, but he doesn't seem to really know what his teacher is talking about. Sam's error indicates that these large numbers he is working with are not very meaningful to him, and so he gets confused more easily than he would if he were working with smaller numbers. It is also apparent that Sam still lacks a complete understanding of conservation of number. This is a concept that Sam has to confront many times before he will really understand it for himself.

Unsure of whether or not it will help Sam, Mr. Bautista decides to ask him another question. "How many do you think there would be in all those tens you are holding if you counted them by ones? How about checking? I'll come back in a minute and see what you found out."

Mr. Bautista makes a mental note to have all the children count in a variety of ways. He knows that this concept depends in part on the development of logical thinking and so it can't be taught directly. However, his job is to ask the children to check any theories they come up with and to challenge them if they have any misconception that he can help clear up. He will model counting by fives and tens for a while in various situations. This will help Sam and others who may be confused about what seems like a simple idea but is really quite difficult for young children to understand.

About the Activities in Section B

Developing Flexibility with Tens and Ones Through Teacher-Directed Activities There are both teacher-directed and independent activities in this section. The independent activities are very important to the development of place-value concepts and should be experienced by the children for several weeks. The amount of time children will need to work with the teacher-directed activities will vary widely. Watch your children closely to see how they deal with the questions you pose. Some will need fairly brief periods of time working with you before they understand the concepts. Others will need continual review of these ideas, even after they have been introduced to addition and subtraction of larger numbers. These activities can be used to develop two ideas simultaneously. First, children explore lots of ways to take numbers apart and put them back together. This helps them see that numbers can be organized in many different ways. For example, 38 can be thought of as 38 ones, as 2 tens and 18 ones, or as 3 tens and 8 ones. This idea is very important to the understanding of addition and subtraction of two-digit numbers.

At the same time, the children are encountering the idea that breaking up a number into parts doesn't change the quantity that the number represents. While this may seem obvious to us, it is not obvious to young children. We often assume that because children have developed an understanding of conservation of small numbers that they also can conserve large numbers. However, this is not the case. This idea develops gradually with experience and maturity and will not be learned by all children in just a few lessons.

In presenting these activities, we must take care not to teach children simply to give the right response. We want to present them with ideas to ponder and to give them opportunities to check their thinking. Teaching them merely to say the right words can only interfere with the sense-making process we are trying to encourage.

Counting and Organizing Numbers into Tens and Ones Through Independent Activities The independent activities in Section B need to be experienced by the children for several weeks. Through these activities children can develop a sense of quantity, recognize the value of organizing numbers into tens and ones, and learn to take numbers apart in a variety of ways. These activities can be made more challenging for the children over time by asking them to make comparisons between quantities.

The activities are easily adapted to meet changing needs over time. Initially, children will be focused on estimating, organizing counters into tens and ones, and recording their results. (The children should be allowed to change their estimates as they get more information.)

Some children are reluctant to make an estimate. Being able to change their mind helps them see that it is easier to estimate after you have more information. I use the terms *estimate* and *guess* interchangeably mainly to de-emphasize the need for children to feel that they need to be "right" even though we hope that children's estimates will get closer to the correct answers over time.

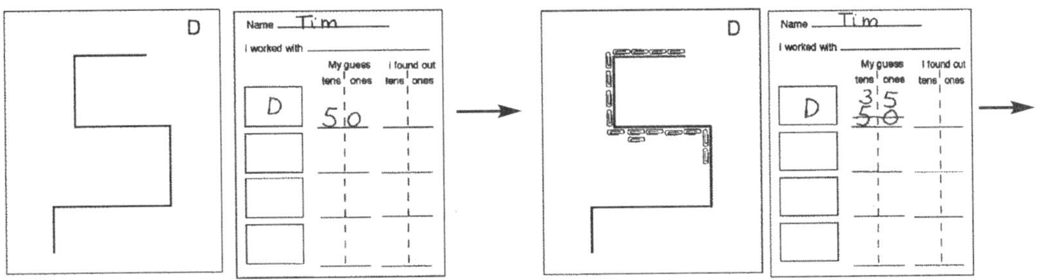

I think 50 paper clips will fit on the line. **I changed my mind. Now I think 35.**

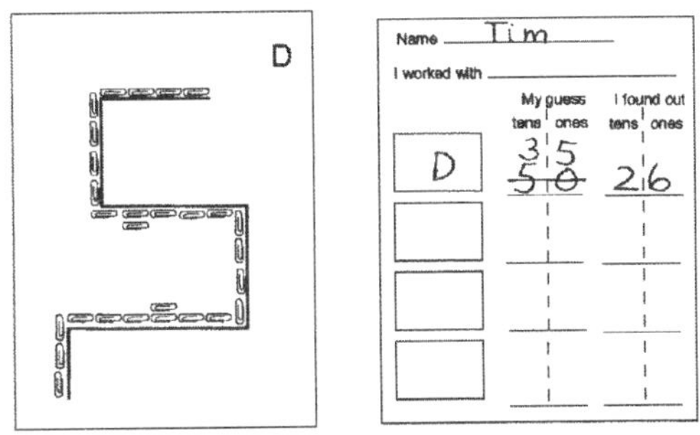

It took 26 paper clips.

Those children who are ready for a challenge can compare quantities. Some will be ready to compare two quantities to see which is more and which is less. Others will be ready to determine how many more or how many less.

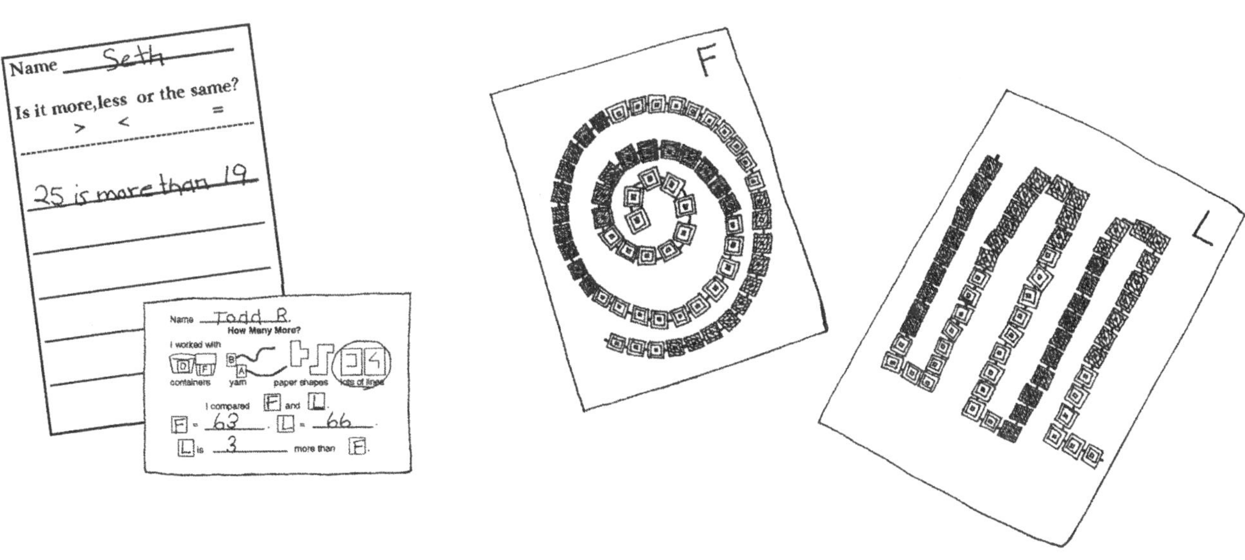

The tasks have been designed to give children experiences with numbers to 100, but they can easily be extended to numbers beyond 100 for children who are ready to work with larger numbers.

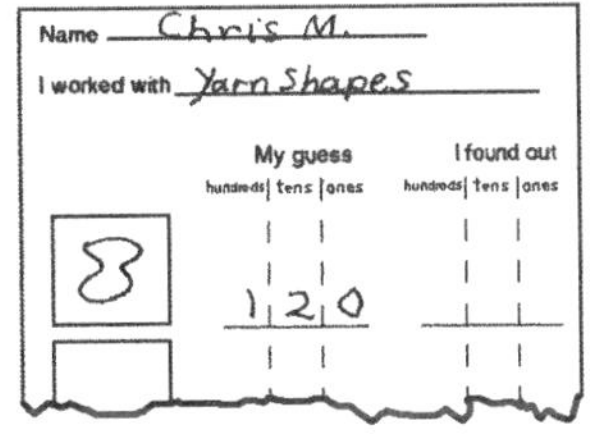

Teacher-Directed Activities

About the Rearrange-It Activities

It seems obvious to us that the number 15 is made up of 1 ten and 5 ones, that 25 represents one ten and 15 ones, and that 36 can be thought of as one ten and 26 ones or 2 tens and 16 ones. But this is not obvious to children. The following activities will help them discover these relationships and develop flexibility in working with numbers.

Do not be surprised if children count and recount as they explore what happens when they arrange and rearrange numbers in several different ways. Continue to help them explore numbers using the activities until these relationships are obvious to them.

1–24 Rearrange-It: Arranging Loose Counters into Tens and Ones

Whole-Class or Small-Group Activity

Materials: Connecting cubes or beans and cups ▪ Place-value boards (1 per child)

In this activity, the children begin with a pile of counters from which they make as many tens as possible.

Using Connecting Cubes

Have the children start by taking a given number of cubes. Say, for example:

Make a pile of twenty-six cubes. Check your friend's pile of cubes while your friend checks yours. Make sure you each have twenty-six cubes.

Snap together ten of your cubes. How many tens do you have?

One ten.

How many leftovers?

Sixteen.

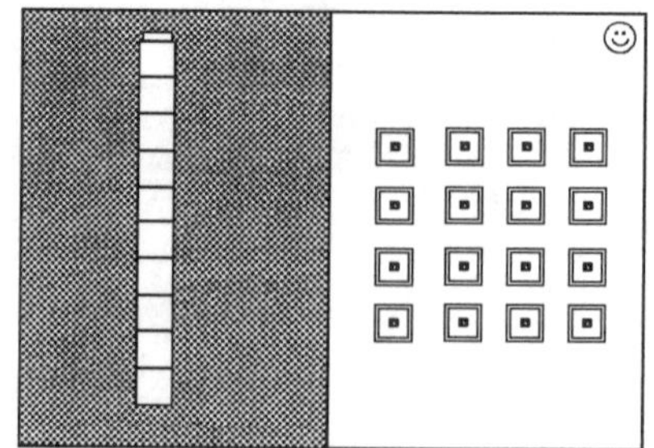

Raise your hand when you can tell me how many cubes you have altogether.

As surprising as it may seem, many children will not realize that they still have the original twenty-six cubes. By having the children raise their hands rather than shouting out their answers, you can determine their various levels of thinking. You will see some children who raise their hands instantly, confident that they have twenty-six cubes, others who will start counting on from ten, and still others who need to count each of the cubes one by one.

Do you have enough loose cubes to make another ten?

Yes.

I don't know.

Let's try to make another ten. Now how many tens do you have?

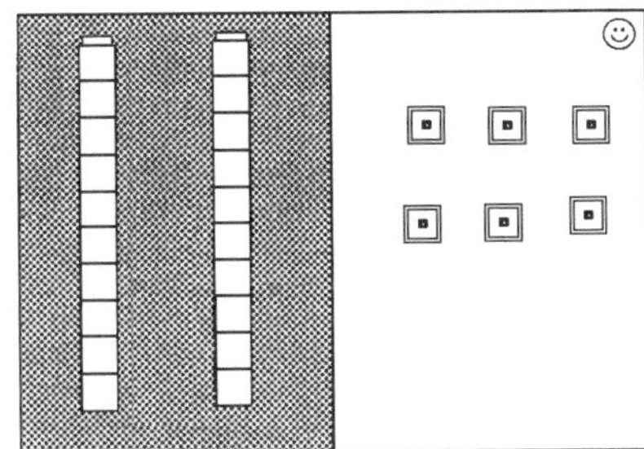

Two tens.

How many leftovers?

Six.

How many altogether?

Allow the children to determine the answer to the last question in any way that makes sense to them, even if they count all twenty-six cubes one by one.

Using Beans and Cups

Using these materials, the activity should proceed in much the same way.

Make a pile of forty-six beans. Check your friend's pile while your friend checks yours. Then put ten beans into a cup.

How many tens do you have?

How many leftovers?

How many altogether?

Do you have enough to make another ten? Try and see.

Now how many tens do you have?

How many leftovers?

How many altogether?

Repeat until all possible tens are made.

1–25 Rearrange-It: Breaking Up Trains into Tens and Ones

Whole-Class or Small-Group Activity

Materials: Connecting cubes or beans and cups ▪ Place-value boards (1 per child)

In this activity, the children start with a long train of connecting cubes and see how many tens they can make from it.

Make a train that is thirty-two cubes long. Line it up with your neighbor's train so that you can make sure that you both have the same number of cubes. How many tens do you think you will be able to make from your train?

> **Three.**
>
> **Four.**
>
> **Two.**

Break up your train into all the tens you can. What do you have?

> **Three tens and two cubes left.**

How many cubes do you have altogether?

Watch to see how the children determine their responses.

Repeat, using other numbers.

1–26 Rearrange-It: Finding All the Ways

Whole-Class or Small-Group Activity

Materials: Connecting cubes (some snapped into trains of ten) ▪ Place-value boards (1 per child)

In this activity, the children learn to recognize that a single quantity can be represented in a variety of ways. Then they try to find all the possible ways. For example:

Get thirty-seven cubes.

Take note of which children count out thirty-seven individual cubes and which get three tens and seven ones.

How can you arrange these cubes on your place-value boards?

> **We could put three tens and seven ones on our boards.**

Let's try that.

Did that idea work? How many cubes do you have altogether?

We still have thirty-seven.

Who has a different idea?

We could just have two tens and the rest loose.

Let's try that. How many loose ones?

Seventeen are loose.

Is there any other way we could arrange them?

They could all be loose.

Repeat, using different numbers.

NOTE: At first, it may seem that having the children arrange their cubes in a variety of ways will distract them from seeing that a particular number is composed of a certain number of tens and ones. However, you want them to be able to think flexibly about numbers and to be able to take them apart in many ways. Later, they will find ways to arrange their cubes most efficiently. (See, for example, Build It Fast, activity 1–29.)

Extension: Interpreting Symbols

Write symbols to describe the arrangements as the children model them. For example:

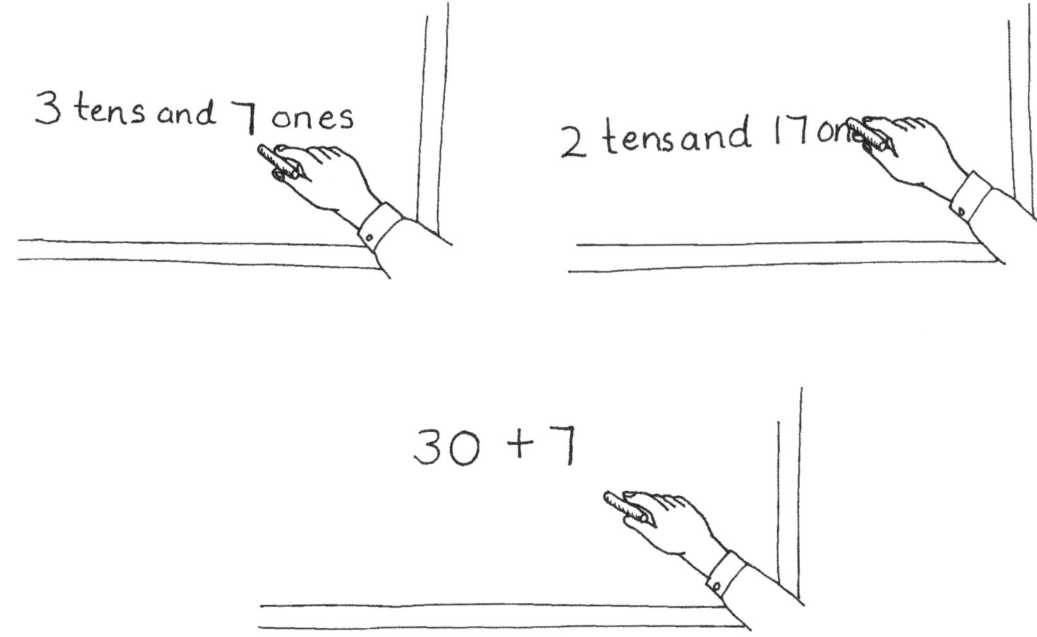

1–27 Rearrange-It: How Many Cubes?

Whole-Class or Small-Group Activity

Materials: Connecting cubes ▪ Place-value boards (1 per child)

With 10 to 20 Cubes

In this activity, the emphasis is on predicting the number of tens and ones that can be made with the "teen" numbers. Direct the children to take a certain number of cubes. For example:

Make a pile of fourteen cubes. Who can predict how many tens you can make? How many ones do you think you will have left? Let's try it and see.

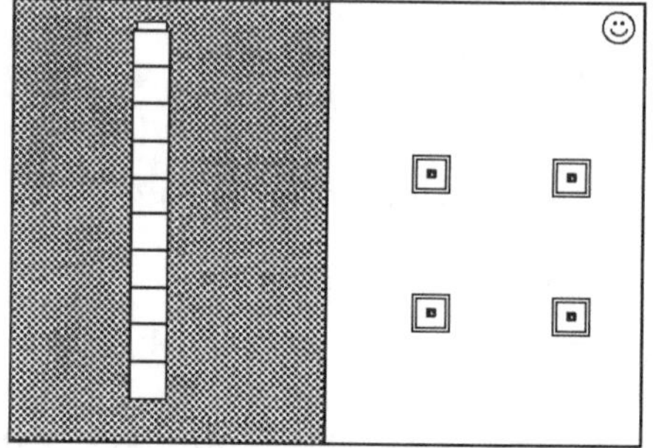

Repeat, using other numbers.

NOTE: Many children still will not be sure that they can make one ten and four ones from fourteen. It will take many experiences like this one before children internalize this concept.

With More Than 20 Cubes

In this activity, the children add numbers from 10 to 20 to groups of ten, giving them another opportunity to think of the "teen" numbers as "tens and leftovers." Have the children start with a larger number of cubes. For example:

Put one ten and fifteen ones on your board. How many cubes do you think you have altogether? Check and see.

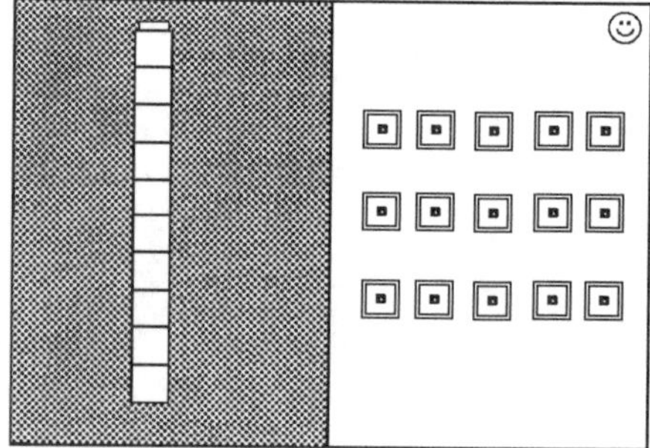

Now clear your board.

Put two tens and thirteen ones on your board. How many cubes do you think you have altogether?

NOTE: Some children may still be counting by ones. Allow those who found the total number of cubes quickly to share with the others how they found it.

1–28 Rearrange-It: Breaking Up Tens .... Whole-Class or Small-Group Activity

Materials: Connecting cubes ▪ Place-value boards (1 per child)

In this activity, children explore what happens to the total number of cubes when they break tens apart. Children need to understand that their total number of cubes does not change even when they break up their trains of ten.

Put three tens and four ones on your board. How many cubes is that?

Thirty-four.

Break up one ten and put the loose cubes with the ones. How many tens do you have now? How many ones? How many altogether?

Repeat, using a variety of numbers.

1–29 Build It Fast .... Whole-Class or Small-Group Activity

Materials: Connecting cubes (some snapped into trains of ten)

In the previous five activities, the children learned to be flexible in arranging cubes in a variety of ways in order to develop the idea of conservation of number. This activity encourages children to arrange cubes in the most efficient manner, helping them develop an appreciation of the usefulness of grouping by tens.

Present a variety of numbers to the children both verbally and in written form. Have them use cubes to build the numbers as fast as they can. Repeat this periodically until the children easily and automatically build numbers as tens and ones in other activities such as those presented in Section C: Addition and Subtraction of Two-Digit Numbers.

1–30 Give-and-Take with Tens and Ones

Whole-Class or Small-Group Activity

Materials: Connecting cubes ▪ 12" × 18" Sheet of paper (for teacher use) ▪ Place-value board (for teacher use)

Put out some cubes on a place-value board so that the children can see how many there are.

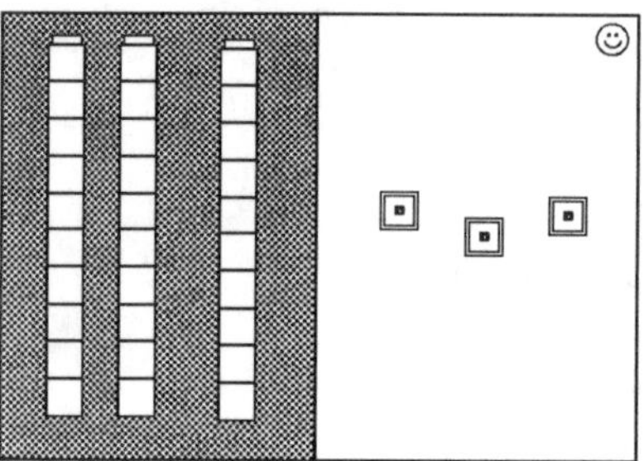

Cover the cubes with a 12" × 18" sheet of paper. Add or remove a few cubes, reaching under the paper and showing and telling the children how many you have added or removed. Do this again and again, each time having the children tell how many they think there are. For example:

I put one more ten on the board. How many do you think there are now?

Lift the paper to enable the children to check.

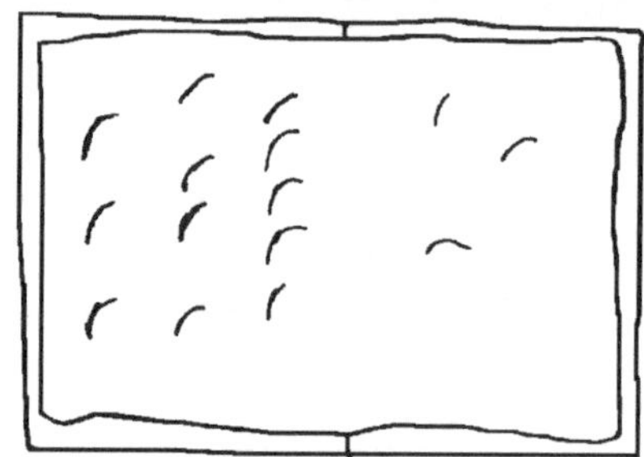

I'm taking one cube off. How many do you think there are now?

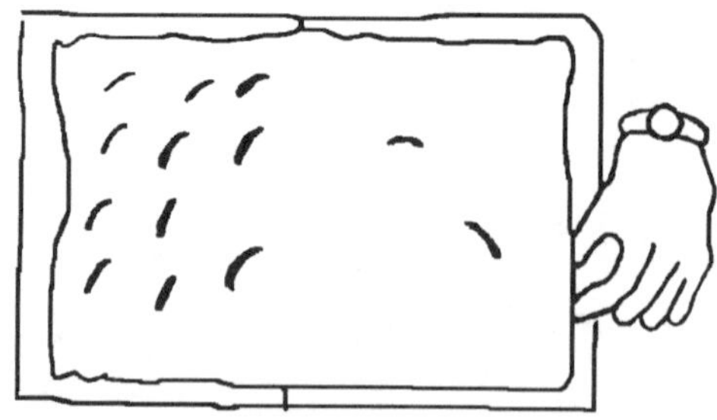

Lift the paper to enable the children to check.

Extension: Writing How Many

Have the children write how many they think are under the paper each time you add cubes or take some away.

1–31 Think About the Symbols

Whole-Class or Small-Group Activity

Materials: Connecting cubes

It is important for children to learn to interpret numerals and to understand how the place in which a numeral is written affects its value.

Occasionally, present children with various ways of expressing numerical amounts so that they learn to think carefully about what the symbols mean. The following list presents examples of ways you can ask children to express the amounts.

40
04
4 tens
4 ones
3 tens and 4 ones
4 ones and 3 tens
1 ten and 0 ones
0 tens and 1 one
5 + 40
40 + 5
30 + 8

Have the children model each of the numbers with connecting cubes to show that they understand the various ways of expressing numerical amounts.

Repeat occasionally until the children can easily interpret the various symbols.

Independent Activities

About Organizing Large Quantities

The following activities give children practice in organizing large quantities into tens and ones. In the beginning, as the children work independently on these activities, allow them to determine the numbers of counters in their own ways. After they have experienced some of the difficulties involved in counting large numbers, talk to them about ways in which they could organize their counters to determine quickly how many tens and ones they have without having to count each one. Tell them that you also want to be able to walk by and quickly see how many counters they have used without having to count each one.

Here are some of the ways the children may find to organize their counters.

- Change colors every time they get to ten.

- Move every tenth cube out of line.

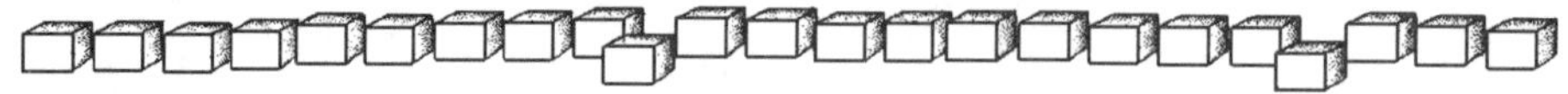

- Mark every tenth counter with another counter in some way.

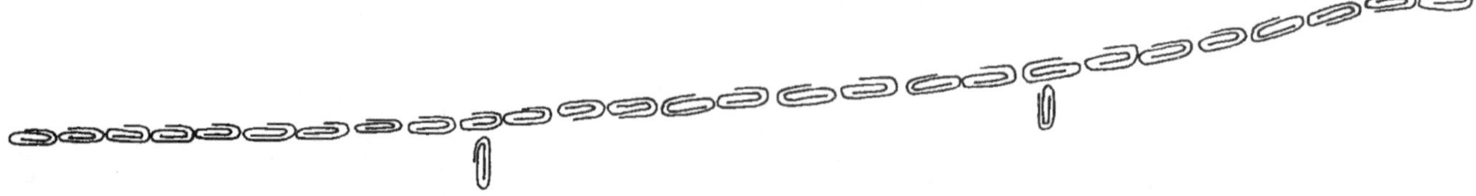

- Use a different color for every tenth cube.

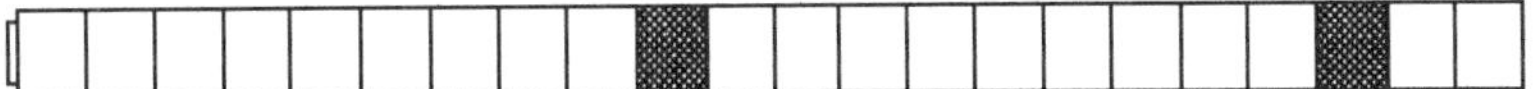

- Snap trains of ten cubes together.

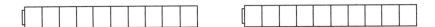

- Group sets of ten together.

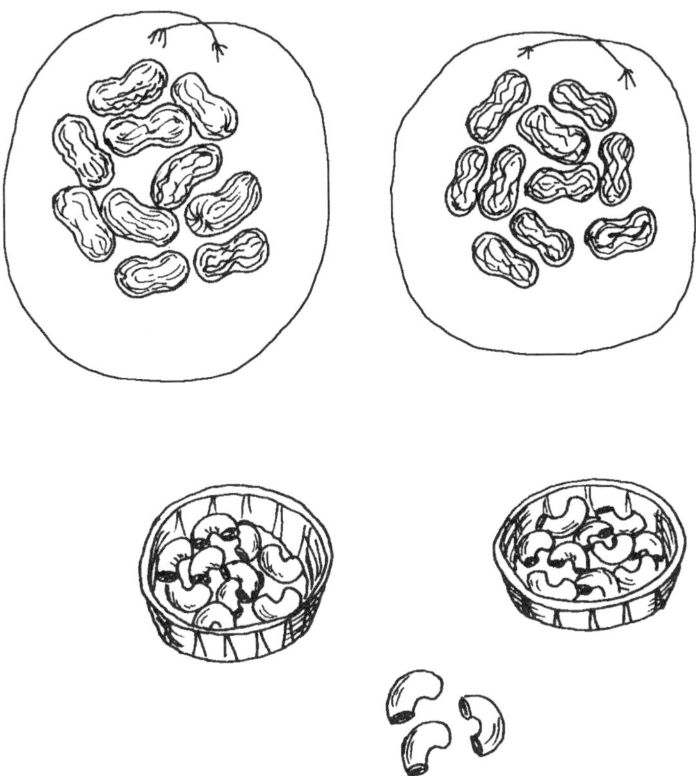

Occasionally, walk by the children while they are working and ask them to tell you how many tens and ones they have used so far. Eventually they will begin to appreciate the organization that prevents them from always having to count by ones.

1–32 Lots of Lines

Independent Activity

Materials: *Level 1:* Connecting cubes (sorted by color), wooden cubes, Color Tiles, or paper clips ▪ Lots-of-Lines task cards (See preparation below.) ▪ Tens and Ones Worksheets (1 per child) [BLM #126]

Level 2: Same as for Level 1, but replace worksheets with More/Less/Same Worksheets [BLM #63].

Level 3: Same as for Level 1, but replace worksheets with How Many More? Worksheets [BLM #128].

Extension: Same as for Level 1, but replace worksheets with Hundreds, Tens, and Ones Worksheets [BLM #127].

Preparation: Make Lots-of-Lines task cards by drawing various line configurations on 9" × 12" pieces of tagboard. Label each card with a letter. (See examples below.)

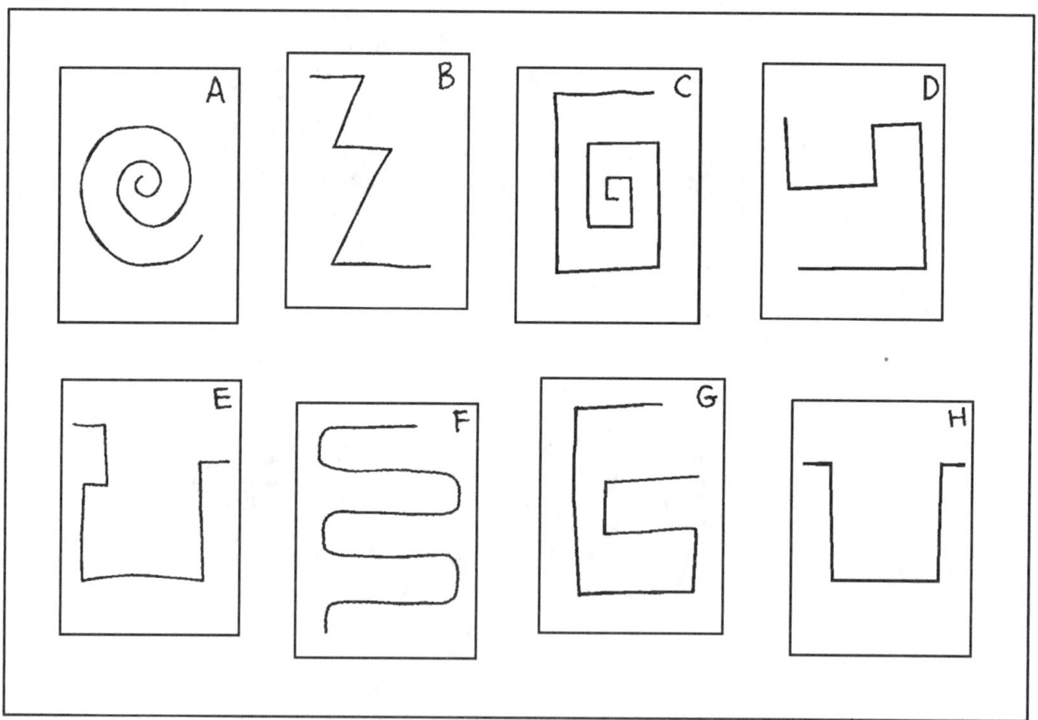

Level 1: Determining a Quantity

The Tens and Ones worksheet is designed for use with several different activities. To complete the sentence at the top of the sheet, "I worked with ______," the children should write "Lots of Lines." If necessary, provide children with this label as a model they can copy.

The children estimate and then figure out how many counters will fit along the lines on each task card. They start by writing the letter of each task card they use in a box on the worksheet. Then they write their estimate, or "guess," and what they found out.

Connecting cubes can be used loose or snapped together, depending on the configuration of the line on the task card.

Level 2: **Comparing Quantities**

The children choose two Lots-of-Lines task cards and measure the lines with counters to determine which card needed more and which needed less. They record their results on the More/Less/Same worksheets using either words or symbols. For example:

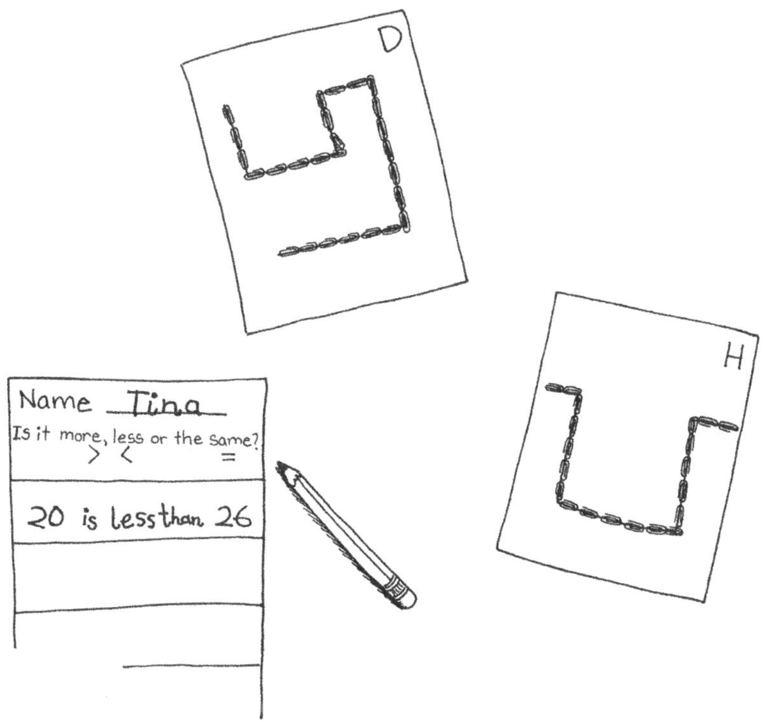

Variation: The children choose three cards. After putting counters along the lines, they order the cards, from the one that holds the least to the one that holds the most. Then they record their results.

Level 3: How Many More or Less?

The children choose two Lots-of-Lines task cards. They measure the lines with counters to determine the length of each line and to find out how many counters longer one line is than the other. Then they record their findings on the How Many More? worksheets.

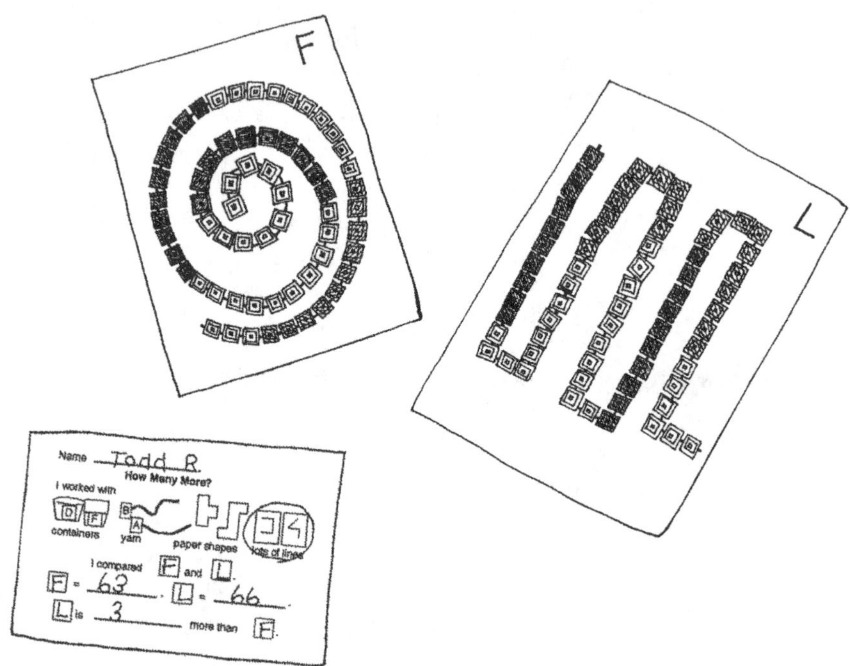

Extension: Larger Numbers

If your children are ready to work with large numbers, provide them with longer line configurations or with smaller counters, such as beans or buttons. As necessary, provide copies of the Hundreds, Tens, and Ones worksheets (so they can work with numbers beyond 100).

1–33 Paper Shapes .... Independent Activity

Materials: *Level 1:* Connecting cubes, wooden cubes, or Color Tiles ▪ Paper Shapes task cards (See preparation below.) ▪ Tens and Ones Worksheets (1 per child) [BLM #126]
Level 2: Same as for Level 1, but replace worksheets with More/Less/Same Worksheets [BLM #63].
Level 3: Same as for Level 1, but replace worksheets with How Many More? Worksheets [BLM #128].
Extension: Same as for Level 1, but replace worksheets with Hundreds, Tens, and Ones Worksheets [BLM #127].
Preparation: Cut shapes of various configurations out of tagboard. Label each with a letter. (See examples below.)

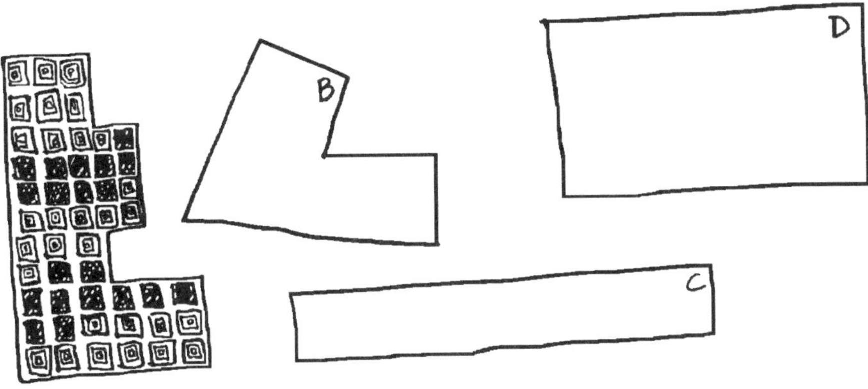

Level 1: Determining a Quantity

The children complete the sentence at the top of the Tens and Ones worksheet with the name of the activity, "Paper Shapes." If necessary, provide cards with this label as a model they can copy.

The children choose a paper shape, write its letter in a box on the worksheet, guess the number of cubes it will take to fill that shape, and record that number. Then they fill in the shape and determine how many cubes they actually needed.

Name Chistopher
I worked with paper Shapes

	My guess tens	My guess ones	I found out tens	I found out ones
A	6	5	5	0

Variation: The children can cut out their own paper shapes and label them with their initials.

Level 2: Comparing Quantities

The children choose two paper shapes and fill them with cubes to determine which holds more and which holds less. They record their results on the More/Less/Same worksheet.

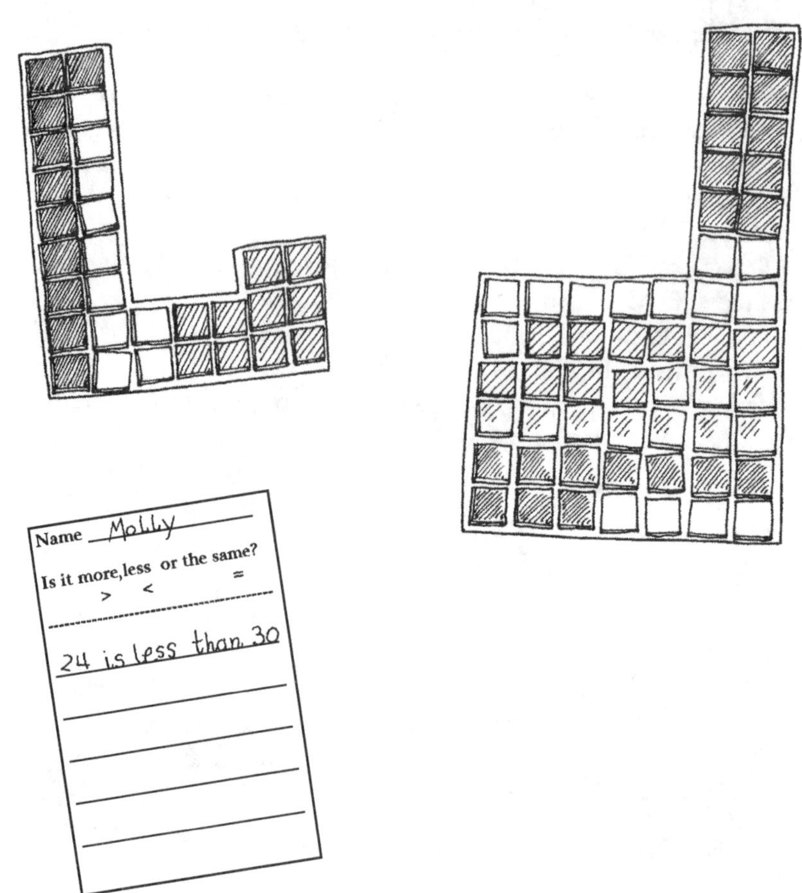

Name Molly
Is it more, less or the same?
> < =
24 is less than 30

Variation: The children choose three paper shapes and determine how many cubes it takes to fill each. They put the shapes in order, from the one that holds the least to the one that holds the most, and record their results.

Level 3: How Many More?

The children choose two paper shapes and fill them with cubes to determine how many more one holds than the other. They record their results on the How Many More? worksheets.

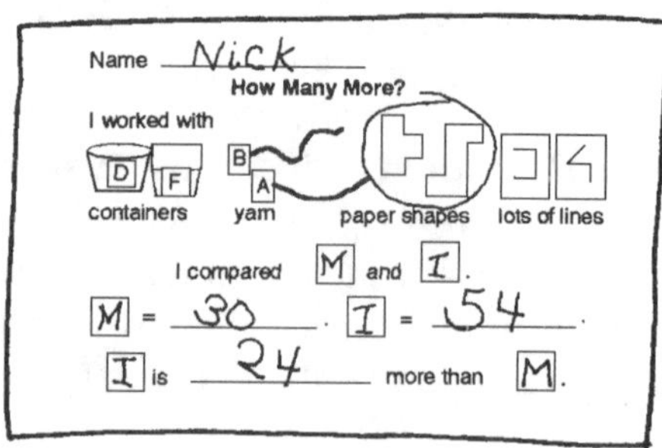

Name Nick
How Many More?
I worked with
D F
B A
containers
yarn
paper shapes
lots of lines
I compared M and I.
M = 30. I = 54.
I is 24 more than M.

Extension: Larger Numbers

If your children are ready to work with numbers beyond 100, provide them with larger shapes or with smaller counters, such as centimeter cubes, and supply the Hundreds, Tens, and Ones worksheets for use as needed.

1–34 Yarn .... Independent Activity

Materials: *Level 1:* Counters ▪ Yarn ▪ Tens and Ones Worksheets [BLM #126]
Level 2: Same as for Level 1, but replace worksheets with More/Less/Same Worksheets [BLM #63].
Level 3: Same as for Level 1, but replace worksheets with How Many More? Worksheets [BLM #128].
Extension: Same as for Level 1, but replace worksheets with Hundreds, Tens, and Ones Worksheets [BLM #127].
Preparation: Cut yarn into various lengths and label each length with a letter written on masking tape.

Level 1: Determining a Quantity

The children complete the sentence at the top of the Tens and Ones worksheet with the title of the activity, "Yarn."

The children first estimate the length of a chosen piece of yarn and then use counters to check their estimates. They record their results on the worksheet.

Level 2: **Quantities**

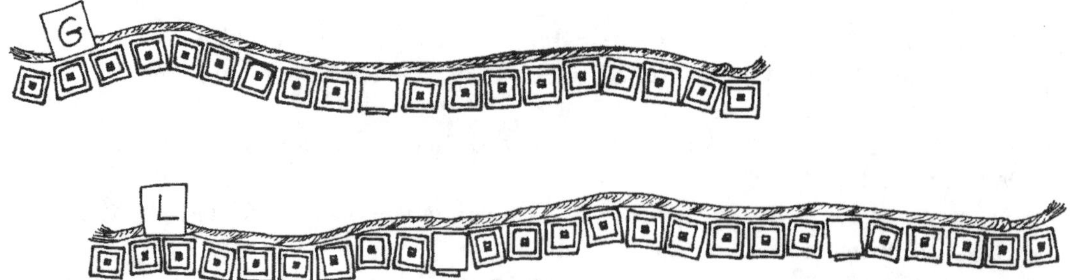

The children choose two lengths of yarn, measure them with counters, and record the results on a More/Less/Same worksheet.

Of course, the children could easily compare the yarn lengths without actually measuring, but in this situation they are to compare the *numbers* they find by measuring the two lengths.

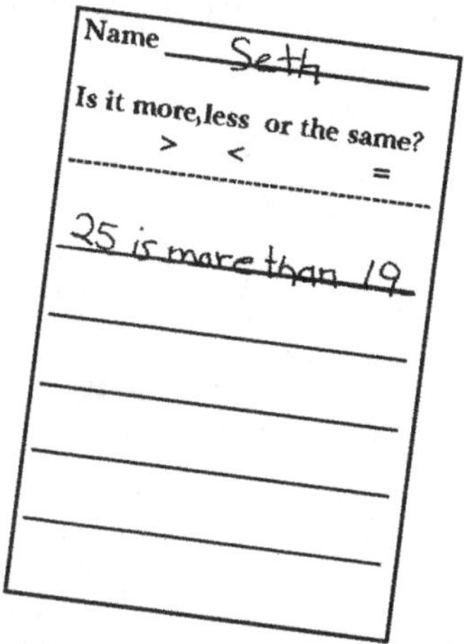
Name Seth

Is it more, less or the same?
> < =

25 is more than 19

Level 3: **How Many More or Less?**

The children compare two pieces of yarn to determine how many more counters long one piece of yarn is than the other. They record their findings on a How Many More? worksheet.

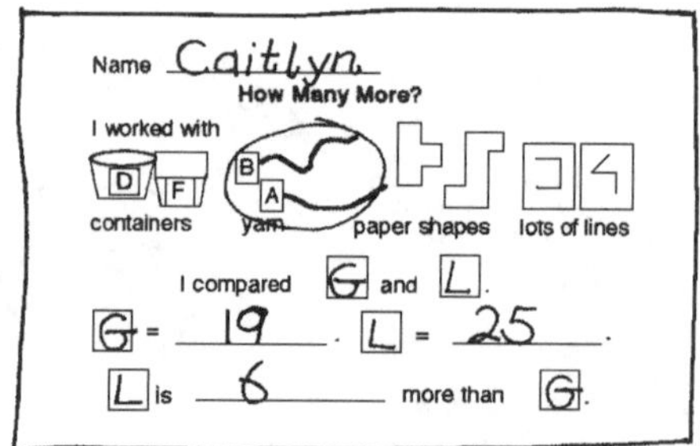
Name Caitlyn

How Many More?

I worked with
containers yarn paper shapes lots of lines

I compared G and L.

G = 19 . L = 25 .

L is 6 more than G.

Extension: Large Numbers

If your children are ready to work with larger numbers, provide longer pieces of yarn or smaller counters, such as beans. Provide the Hundreds, Tens, and Ones worksheet as necessary.

1–35 Yarn Shapes .... Independent Activity

Materials: Lengths of yarn labeled with letters written on masking tape (as prepared for activity 1–34) ▪ Connecting cubes or Color Tiles (sorted by color) ▪ Tens and Ones Worksheets [BLM #126] or Hundreds, Tens, and Ones Worksheets [BLM #127]

The children fill in the top of the Tens and Ones worksheet with "Yarn Shapes," and write the letter of the piece of yarn they will be using. They form the yarn into a closed shape. Then they draw this shape in the first box on the worksheet.

Children estimate the number of cubes or tiles it will take to fill in their yarn shape. They record their estimate on their worksheet. They then fill in the shape with cubes or tiles to find out how many it actually takes and record that number.

I'll make this shape with my yarn.
I think it will hold seventy-five cubes.

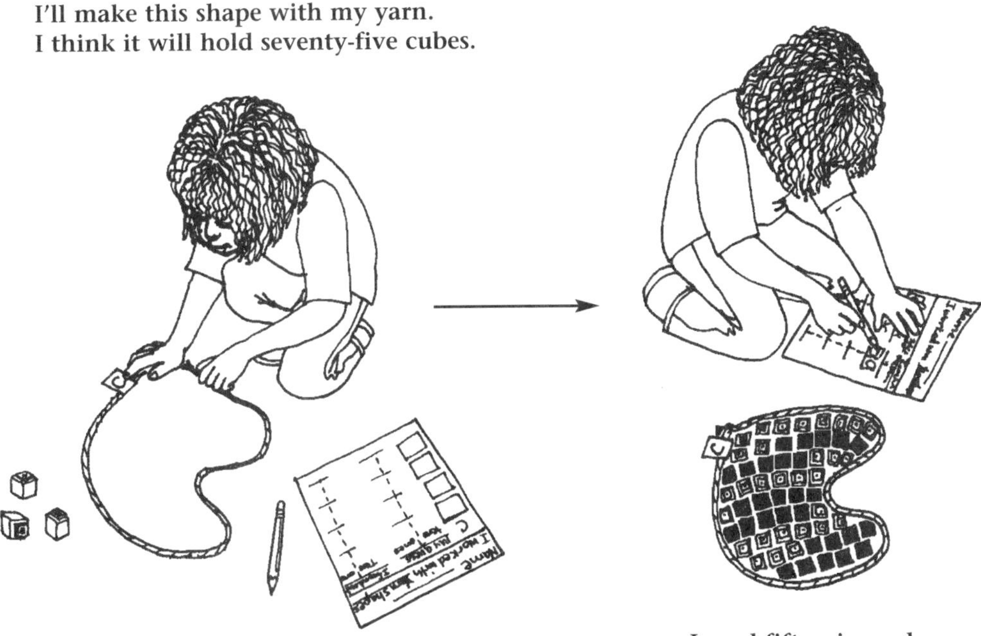

I used fifty-nine cubes.

Using the same piece of yarn, children make a different shape, which they then draw in the next box. They make their guess, fill the shape with counters, and write how many they used. They should use the same piece of yarn to form and measure at least four different shapes. For example:

Extension: Comparing Shapes

Some children will be ready to write about what they noticed about their four shapes. Suggest that they tell which shape held the most, which held the least, and why.

1–36 Containers

Independent Activity

Materials: *Level 1:* Counters • Various empty containers (margarine tubs, cans, jars, small boxes) labeled with letters • Tens and Ones Worksheets [BLM #126]
Level 2: Same as for Level 1, but replace worksheets with More/Less/Same Worksheets [BLM #63].
Level 3: Same as for Level 1, but replace worksheets with How Many More? Worksheets [BLM #128].
Extension: Same as for Level 1, but replace worksheets with Hundreds, Tens, and Ones Worksheets [BLM #127].

Level 1: Determining a Quantity

Children complete the sentence at the top of their Tens and Ones worksheet with the name of the activity, "Containers." (Provide a model if necessary.)

The children choose a container and estimate the number of counters it will take to fill it. Then they fill it to determine how many it will hold and record their results on their worksheet.

Level 2: Comparing Quantities

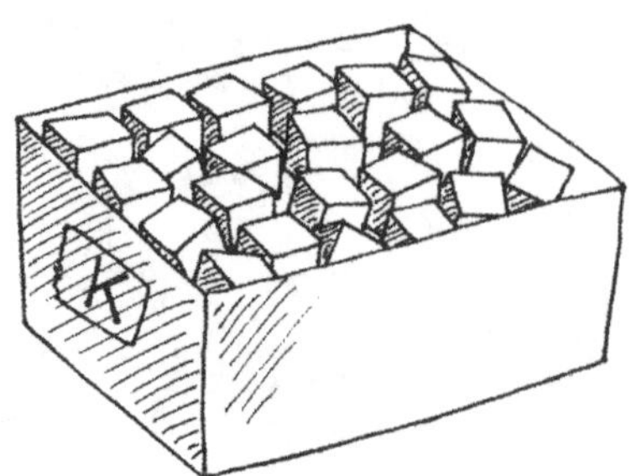

The children choose two containers and find out how many counters each holds. They record their results on a More/Less/Same worksheet, using either words of symbols.

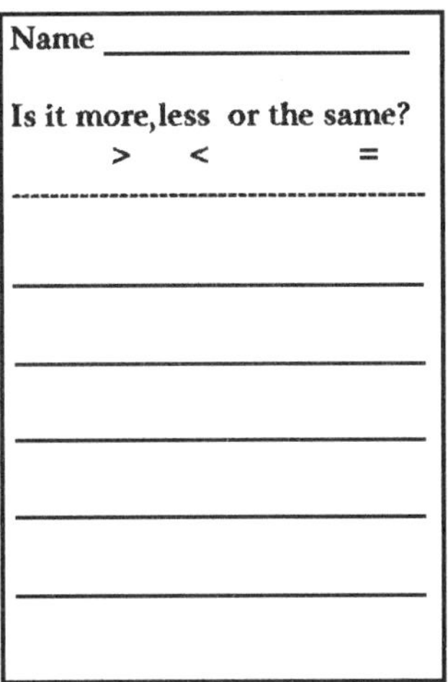

Name ________________

Is it more, less or the same?
> < =

Variation: The children choose three containers and determine how many each holds. They order the three, from the one that holds the least to the one that holds the most, and record the results on a worksheet.

Level 3: **How Many More?**

The children choose two containers and compare them to see how many more counters one holds than the other. They record their results on a How Many More? worksheet.

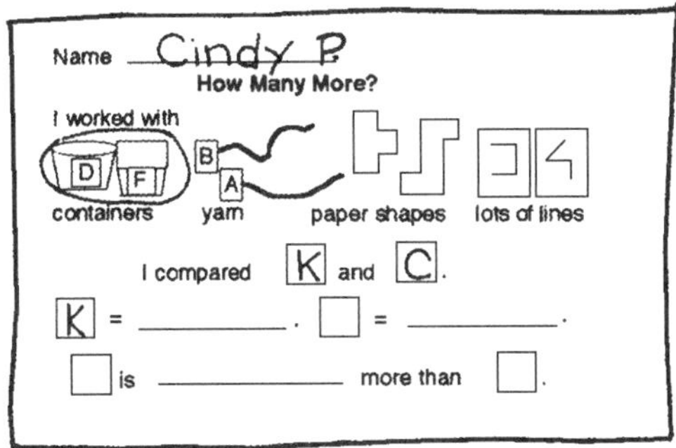

Name Cindy P.

How Many More?

I worked with

containers yarn paper shapes lots of lines

I compared K and C.

K = ________. ☐ = ________.

☐ is ________ more than ☐.

Extension: Larger Numbers

If your children are ready to work with numbers beyond 100, provide them with larger containers or with smaller counters. Make available the Hundreds, Tens, and Ones worksheets.

1–37 Cover It Up

Independent Activity

Materials: *Level 1:* Counters ▪ A list of things in the room with a measurable surface area; for example, a book, a child's desktop, a chair seat, an outline of a child's foot, the bottom of the wastebasket ▪ Tens and Ones Worksheets (1 per child) [BLM #126]
Level 2: Same as for Level 1, but replace worksheets with More/Less/Same Worksheets [BLM #63].
Level 3: Same as for Level 1, but replace worksheets with How Many More? Worksheets [BLM #128].

Level 1: Determining a Quantity

Children complete the sentence at the top of their Tens and Ones worksheet with the name of the activity, "Cover It Up." (Provide a model if necessary.) The children choose something from the list of things in the room, determine the number of counters needed to cover its surface, first guessing, then using counters, and record their results on the worksheet. For example:

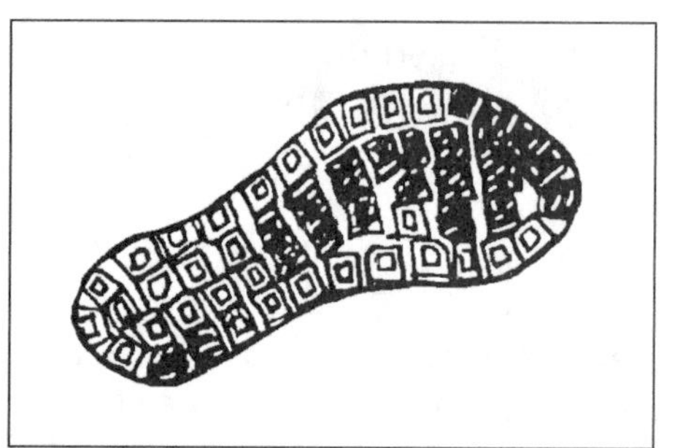

Encourage the children to suggest other things to measure. Add their suggestions to your list of possibilities.

Level 2: Comparing Quantities

The children choose two things to measure and compare them to see which has the larger area and which has the smaller. They record their results on a More/Less/Same worksheet.

Variation: The children choose three things to cover, order them by the number of counters they used to cover them, and record their findings.

Level 3: How Many More?

The children choose two things to cover and measure them to determine how many more counters one needs than the other. They record their results on a More/Less/Same worksheet.

1–38 Measuring Things in the Room* .... Independent Activity

Materials: Counters ▪ Measuring Things cards (See preparation below.) ▪ Tens and Ones Worksheets [BLM #126]
Level 2: Same as for Level 1, but replace worksheets with More/Less/Same Worksheets [BLM #63].
Level 3: Same as for Level 1, but replace worksheets with How Many More? Worksheets [BLM #128].
Preparation: To make Measuring Things cards, sketch things in the room that the children can measure, one to a card. If you don't want the children to work with numbers greater than 100, be sure not to include objects that are more than 100 counters long.

* Based on *Mathematics Their Way,* "Measuring," p. 307.

Level 1: Determining a Quantity

Children complete the sentence at the top of the Tens and Ones worksheet with the name of the activity, "Measuring Things." (Provide a model if necessary.) Each child chooses a Measuring Things card, locates the actual object in the room, and guesses its length in terms of a number of counters of one kind. After recording the guess on the worksheet, the child measures the length of the object with the counters and records the result.

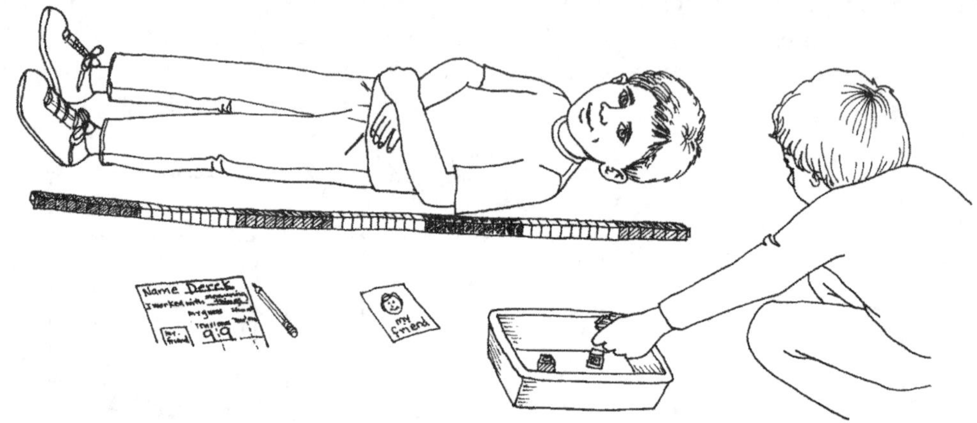

Level 2: Comparing Quantities

The children measure and compare the lengths of the actual things pictured on two cards. They record the results on a More/Less/Same worksheet.

Level 3: How Many More?

The children choose two cards and measure the actual objects pictured to determine how much longer one object is than the other. They record their results on a Measuring Things worksheet.

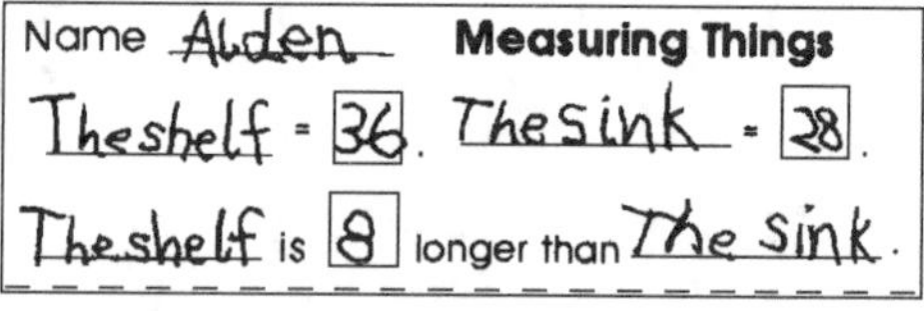

Name Alden **Measuring Things**

The shelf = 36. The sink = 28.

The shelf is 8 longer than The sink.

1–39 Measuring Myself .... Independent Activity

Materials: Connecting cubes ▪ Measuring Myself cards (see below) ▪ Tens and Ones Worksheets [BLM #126]
Level 2: Same as for Level 1, but replace worksheets with More/Less/Same Worksheets [BLM #63].
Level 3: Same as for Level 1, but replace worksheets with How Many More? Worksheets [BLM #128] ▪ Measuring Myself Worksheets [BLM #131]
Preparation: To make Measuring Myself cards, sketch stick figures, each with a highlighted body part on one side of the card. Write words that describe the body part on the other side, as shown in the examples below.

Level 1: **Determining a Quantity**

Children complete the sentence at the top of the Tens and Ones worksheet with the name of the activity, "Measuring Myself." (Provide a model if necessary.) Each child chooses a Measuring Myself card and then decides how to measure the body part with connecting cubes. Because some body parts are difficult to measure on oneself, children may need to work with partners.

The children estimate, measure, and record their results on the Tens and Ones worksheet.

Name Taylor

I worked with measuringmyself

	My guess tens	My guess ones	I found out tens	I found out ones
B	5	0		

Level 2: **Comparing Quantities**

The children choose two cards and measure the body parts. They determine which is longer or shorter and record their results on the More/Less/Same worksheet.

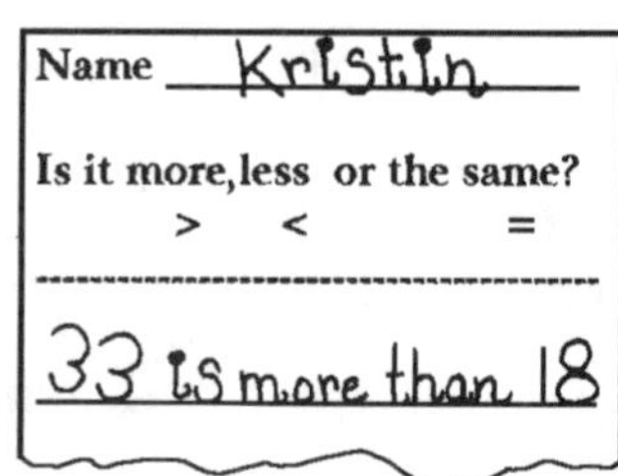

Name Kristin

Is it more, less or the same?

> < =

33 is more than 18

Level 3: **How Many More?**

The children measure and compare two body parts and record how much longer one is than another on the Measuring Myself worksheet.

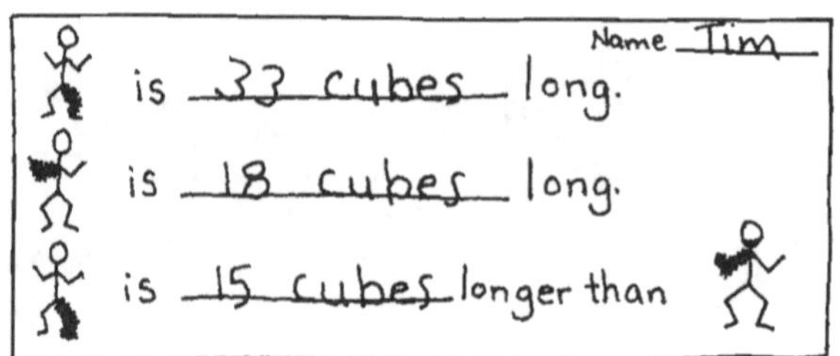

Name Tim

is 33 cubes long.

is 18 cubes long.

is 15 cubes longer than

1–40 Comparing Myself .... Independent Activity

Materials: Connecting cubes ▪ Measuring Things cards (as prepared for activity 1–38) ▪ Compared-to-Me worksheets [BLM #130]

Have the children measure their height using connecting cubes and record it at the top of the Compared-to-Me worksheet.

Then children choose Measuring Things cards and measure the objects to determine their length. They list each thing measured in the appropriate column on the worksheet, writing the length of each and indicating whether it was longer or shorter than they are.

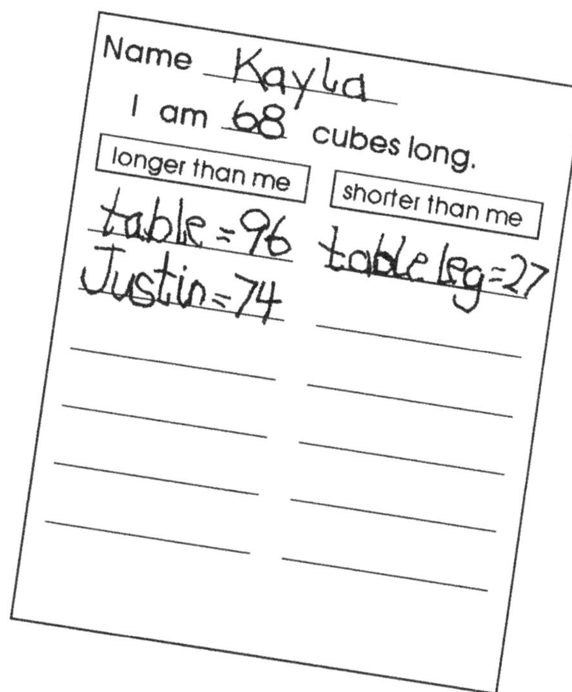

1–41 Making Trails

Independent Activity

Materials: Counters ▪ Make-a-Trail cards (See preparation below.) ▪ Making Trails Worksheets [BLM #132]

Preparation: Make a set of Make-a-Trail cards. Find places in your classroom between which the children can easily measure the distances. Describe each trail and illustrate it on a 3" × 12" tagboard card. Consider the numbers you want the children to work with when creating the cards, keeping in mind that the numbers of counters used to measure the distance between two places can quickly become quite large.

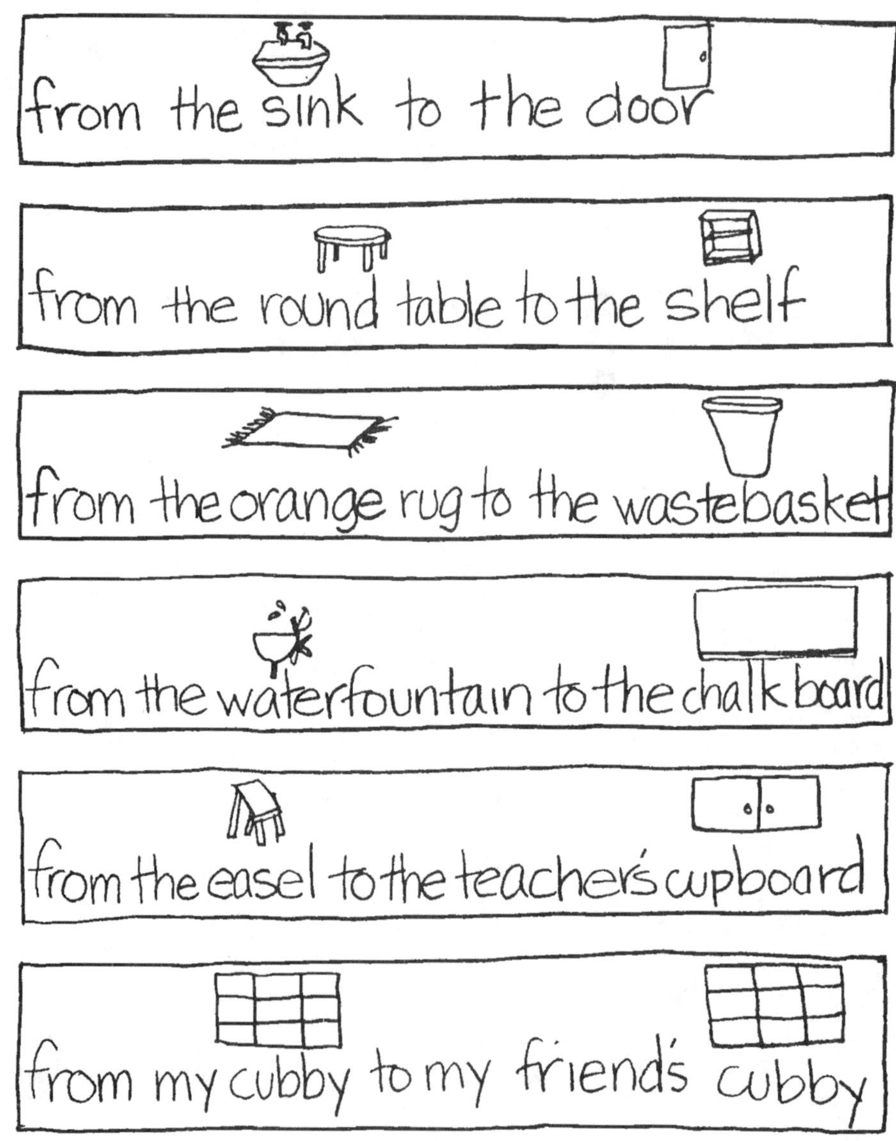

The children choose a Make-a-Trail card and determine the number of counters it takes to make a trail from one place in the room to another, as indicated on the card. For example:

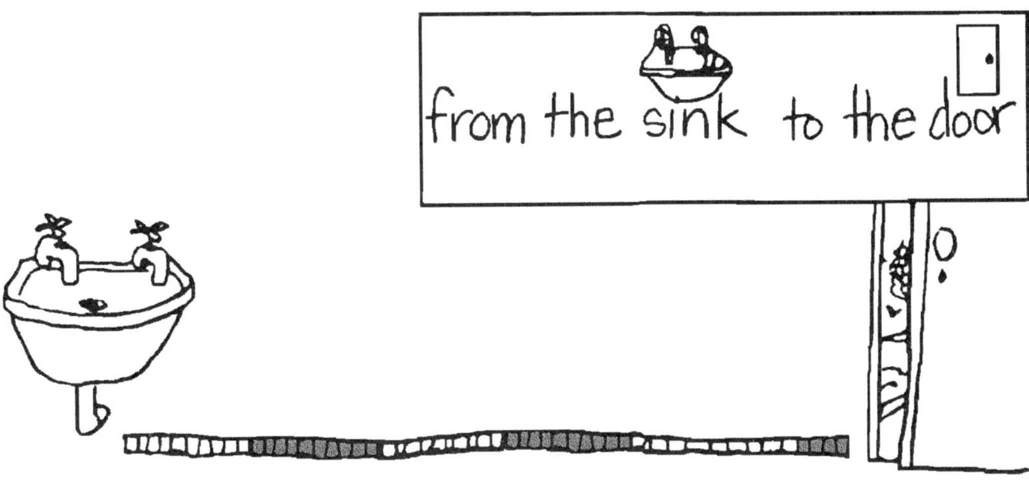

The children then record the number of counters they used by completing the sentences on the worksheet.

I made a trail from
the _____ to the _____.
I guessed _____.
I used __________.

I made a trail from
the sink to the door.
I guessed 40.
I used 53 counters.

1–42 Building Stacks*

Independent Activity for Partners

Materials: Connecting cubes • Build-a-City Game Board (1 per pair) [BLM #32] • Use a More-or-Less Spinner if you have one. (See Book 1, p. 145.) Otherwise, make a More-or-Less die by marking three faces of a wooden cube with "More" and three faces with "Less." • 4–9 Number cubes (1 per pair) • More/Less/Same Worksheets [BLM #63]

Preparation: Duplicate Build-a-City game boards on construction paper or tagboard.

Players place the Build-a-City game board between them. They take turns rolling a number cube and placing the indicated number of connecting cubes in a stack, or "building," on their side of the game board. With each roll, they place another building on a new section of the game board.

* Based on *Mathematics Their Way,* "Unifix Stacks," p. 320.

When all sections of the board have been filled, players each take apart their buildings and count their cubes, first arranging them into tens and ones.

Each player records both numbers of cubes on a More/Less/Same worksheet.

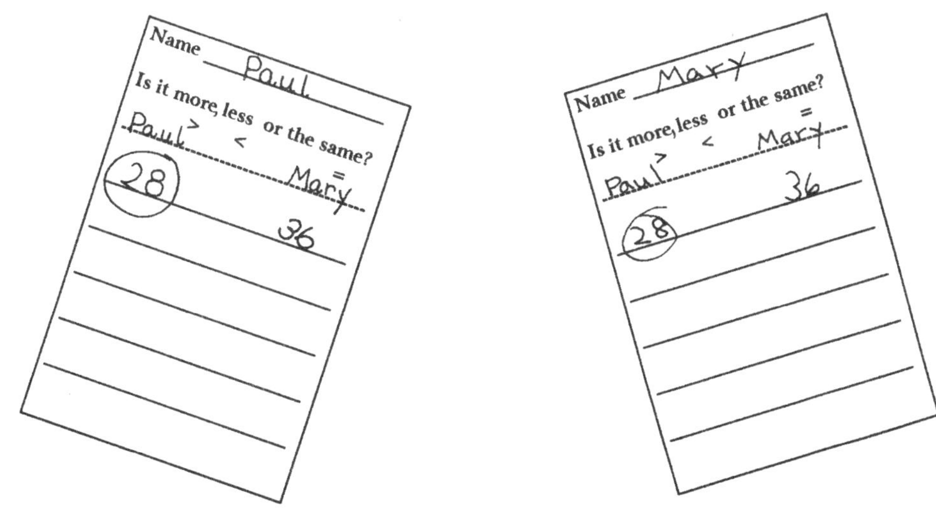

One player spins the More-or-Less Spinner or rolls the More-or-Less cube to find out whether the player with more cubes or the player with less wins this round of the game. The players circle the winning number and play again.

1–43 Race to 100

Independent Activity for Up to Four Children

Materials: Connecting cubes ▪ Place-value boards (1 per child) ▪ 1–6 Number cubes (1 per pair) ▪ Paper plates (each to hold 100 cubes)

Players take turns rolling a number cube to determine how many connecting cubes to place on their place-value boards. The object of the game is to be the first to reach 100.

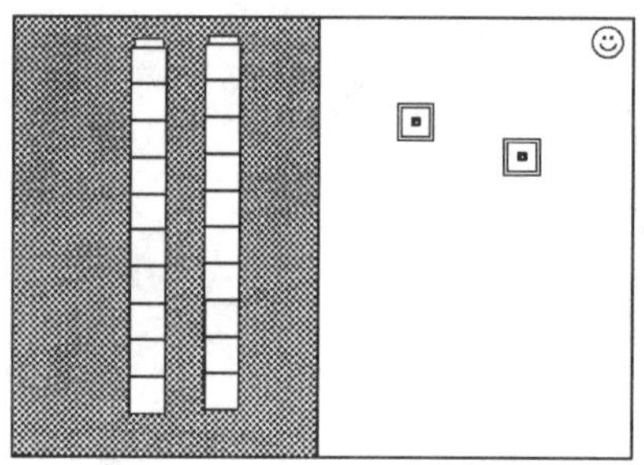

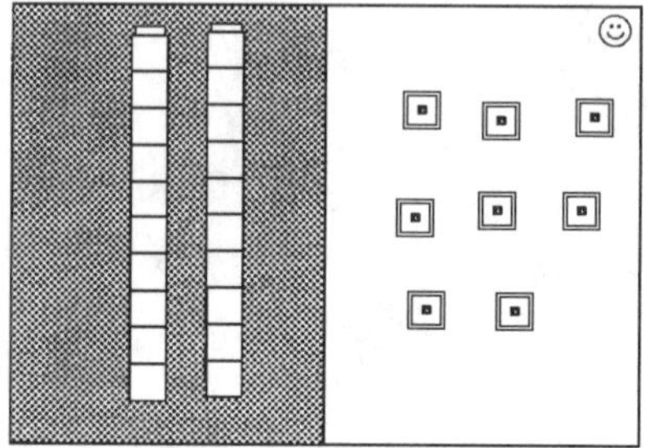

I rolled four. I will have enough for another ten.

There are two possible ways to end the game. Suppose, for example, that a child needs four more cubes to make 100. If the child rolls a five, he or she could win by getting to 100 with one left over. Alternatively, players might establish ahead of time that a player has to reach 100 *exactly* to win the game. In this case, the child who needs four to reach 100 would either have to roll a four or miss a turn.

Variation: Children can play the game so that the winner is the last player to reach 100.

1–44 Race to Zero

Independent Activity for Partners

Materials: Connecting cubes ▪ Place-value boards (1 per child) ▪ Number cubes ▪ Paper plates (each to hold 100 cubes)

Each player starts with 100 cubes. Players take turns rolling a number cube to determine how many connecting cubes to take off their boards. The winner is the first player to reach zero.

Unlike Race to 100 (activity 1–43), there is only one way to win in this game. An exact number must be rolled. So, for example, if a player has four cubes on the board and rolls a five, since it is impossible to take five away from four, the player would miss that turn.

SECTION C

ddition and Subtraction of Two-Digit Numbers

For many years, I believed my job was to teach my students a particular way to add and subtract using regrouping. I thought I was supposed to teach borrowing and carrying following a procedure familiar to most of us. For example:

```
   1             2 1
   47             32   (3 crossed out)
 + 28           - 17
 ----           ----
   75             15
```

This section presents a process that can be used to help children develop competence and efficiency when adding and subtracting numbers to 100 and beyond that is different from the one most of us learned when we were in school and different from what most of us have used to teach children in the past. The process is easy and very effective, but because it is different, I will present the thinking behind the decision to use this process.

Why Do We Need a Different Way? Consider the following problem:

Double 256.

Some people would be most comfortable by first writing the problem and then adding in the way they were taught in school. For example:

We may think,"6 and 6 is 12. Put down the 2 and carry the 1.

```
    1
   256
 + 256
 -----
     2
```

Then we may think, "5 and 5 is 10. Add the 1.That makes 11. Put down the 1 and carry the 1."

```
   1 1
   256
 + 256
 -----
    12
```

Notice that we have gone through several steps and have no way to reflect on whether or not the answer we have so far is reasonable.

```
   1 1
   256
 + 256
 -----
    12
```

Next we add 2 and 2 and 1 more and end up with a 5. When we add this way, we get an answer but there has been no need to think about the numbers or their relationships. Getting a right answer is dependent on following the right steps.

$$\begin{array}{r} {}^{1\,1}\\ 256 \\ +\ 256 \\ \hline 512 \end{array}$$

It is not necessary, however, to follow a particular procedure to get a right answer. Some children who look for relationships when solving problems might think:

Doubling 250 would give us 500. Adding 6 and 6 gives us 12. So the answer is 512.

Others might think this way:

Since 200 and 200 is 400 and 50 and 50 is another 100, that makes 500. Six and 6 is 12, so that makes 512.

Those who see these relationships can arrive at the answer quickly and, at the same time, can see whether or not the answer makes sense.

Contrast this sense-making process with the struggle many children have when they are asked to remember the steps the teacher has taught them. Teachers have seen that children's common errors, such as the following, reveal their lack of understanding.

$$\begin{array}{r} 46 \\ +\ 59 \\ \hline 915 \end{array} \qquad \begin{array}{r} 92 \\ -\ 16 \\ \hline 84 \end{array} \qquad \begin{array}{r} 145 \\ +\ 36 \\ \hline 505 \end{array}$$

When children focus on following the steps taught traditionally, they usually pay no attention to the quantities and don't even consider whether or not their answers make sense. Because such children are just memorizing steps they don't understand, the procedures must be reviewed year after year. The same kinds of mistakes repeatedly crop up each year, the only difference being that children are working with bigger and bigger numbers.

Children who have memorized a procedure often do not use common sense in situations where the procedure may not be the most efficient way to solve the problem. This was evident to me one day when I saw a second grader struggling to solve this addition problem by following a procedure he had memorized.

$$\begin{array}{r} 36 \\ +\ 98 \\ \hline \end{array}$$

If the child had learned to look at relationships, he would have been able to approach the problem by thinking about it, rather than just by following the prescribed steps. So, for example, he might have thought this way:

I want to add 98 to 36. I know that 98 is almost 100 and 100 + 36 is 136. Since 100 is two more than 98, I need to take 2 off. The answer is 134.

While it takes time to explain the thinking behind this solution, the actual solving of the problem is almost instantaneous.

Even solving simpler problems is often not obvious to children who are focused on procedures. We want children to know, for example, that the sum of 10 + 24 must be 34. Instead, many children go through the process of first adding 4 + 0 and getting 4 and then adding 2 + 1 and getting 3. And we have all seen children "borrowing" when solving a problem such as 14 − 7. They carefully cross out the 1 ten and rewrite it next to the 4, ending up with the same problem they started with!

$$\begin{array}{r} 34 \\ +\ 10 \\ \hline \end{array} \qquad\qquad \begin{array}{r} \overset{1}{\not{1}}4 \\ -\ 7 \\ \hline \end{array}$$

Any teacher who has worked with young children has heard them ask—even before they look at their papers—"Is this when we carry?" and "Is this when we borrow?" When we teach children a procedure that stops them from looking at relationships, their focus is naturally on those steps that lead them to the "right answer." And the many children who cannot yet make sense of the procedure are forced to stop trying to understand and to begin working to learn the steps, whether or not they make sense to them.

Of course, some children do see the relationships underlying the steps. Some others memorize the procedures easily. But teachers know that most young children do not learn this easily and very few understand what they are learning. Even those children who are successful at getting right answers are often unable to use these procedures to solve problems. Because they have learned a procedure they do not understand, they cannot apply it in new situations. It's like moving to a new town and learning your way from your home to the store. If you follow the steps exactly, you can get where you are going. But if you take a wrong turn, you're lost. You have no way of using what you do know to help you figure out what you don't know.

We want to give children more powerful and sensible ways for them to add and subtract that allow them to develop both understanding and facility with addition and subtraction.

What Is the Process? What is the Rationale Behind It? This process is based on the premise that children are more mathematically powerful and more capable of adding and subtracting large numbers when they are allowed to work with the numbers in the way that makes the most sense to them. It is easier for children to understand what they are doing when they are looking for relationships between numbers than it is for them to merely follow a prescribed method.

If children are to become proficient in using number relationships to solve problems, it is important for them to work with place-value concepts using manipulatives as presented in Sections A and B. Ultimately, the goal is for children to do addition and subtraction problems without relying on counters. However, children do not start out doing this. They should begin their work with addition and subtraction using physical models that help them see relationships and check their thinking. The process presented here shows you how to use manipulatives to help children learn to think with numbers and solve problems using relationships.

Goals for Children's Learning* *(Section C)*

Goals

Given a variety of problems to solve, the children will:

- Interpret addition and subtraction problems with manipulative materials
- Develop strategies to determine *how many*
- Solve problems in a variety of ways
- Consider whether or not their answers make sense

Analyzing and Assessing Children's Needs

In order to determine what further work the children need with addition and subtraction of two-digit numbers, we must know more than whether or not they can get correct answers to worksheet problems. We must know if they understand the process and can be flexible when solving problems. Too many children can get the right answers but cannot apply this process in real situations. We can find out about our children's understanding of addition and subtraction of two-digit numbers by observing them while they solve problems. Keep in mind the following questions as you watch children working on the place-value activities in Section C.

Questions to Guide Your Observations*

Questions

- When given a two-digit addition problem, can the children interpret it with manipulative materials?
- Can they interpret a two-digit subtraction problem with manipulatives?

* Adapted from *How Do We Know They're Learning? Assessing Math Concepts.*

- What methods do they use to solve the problems?

  Do they count every number? Do they count by twos or fives? by tens?

  Do they count on?

  Do they take numbers apart and combine them with other numbers? Do they try to make tens? What else do they do to determine the answers?

  Do they know any of the basic patterns that are helpful in solving problems, such as adding or subtracting 10, adding or subtracting 20, and adding or subtracting 9?
- Can they interpret word problems and write the appropriate equations? Can they solve the equations with models or without models? Can they demonstrate their solutions with models even if they didn't use models to figure out the solutions?
- Do they consider whether or not their answers make sense?
- Are they willing to share their solutions with confidence or do they hesitate to share them?

Meeting the Range of Needs

As long as everyone has access to manipulative materials, all children should be successful at solving addition and subtraction problems. It is critically important that children be allowed to develop their own strategies for working with these problems. They will have the opportunity to develop more sophisticated methods over time as they hear how other children solve the problems and as they have experiences solving problems themselves.

You will be able to meet a range of needs if you sometimes work with children in small groups. Children will benefit from the opportunity to work in small groups for a couple of reasons. You will be able to more closely match the level of problems with children's particular needs, and you will be able to provide for those children who are more able to listen to others and share their ideas when in a small group. Do not set up permanent groups. Respond to particular needs as they emerge. Sometimes you will want to work with a small group of children with similar needs. Other times you will want the small group to be diverse and able to discuss and share a variety of methods.

A Classroom Scene

Mrs. Patwin's second grade has worked with the place-value activities for several months. The children are very comfortable working with tens and ones in a variety of situations. Mrs. Patwin has been working with them as a whole group practicing addition and subtraction two or three times a week, for about four weeks now, and her children are familiar with the routine.

Whole-Class Activity: *Adding and Subtracting Two-Digit Numbers*

When the class comes in from recess, the children all sit in a circle on the rug. Two children pass out place-value boards and three other empty containers of connecting cubes onto the floor in the middle of the circle. Everyone works to make lots of trains of ten. When it appears that the children have made enough tens to work with, Mrs. Patwin asks for their attention. She writes the number 26 on the chalkboard. "I want you to build this number on your place-value boards," she tells them.

This activity is easy for everyone except the new boy, Stephen, who puts two cubes on the left side of his board and six on the right. He then looks over at the other children and notices that they each used two tens trains on the left instead of two cubes. He fixes his board immediately and watches the other children carefully to see what to do next.

"Now I'm going to write something else on the chalkboard," the teacher says. "I want you to think about what the answer might be before you do anything with your cubes. When you think you have an answer, put your thumb under your chin." Mrs. Patwin writes the following on the board.

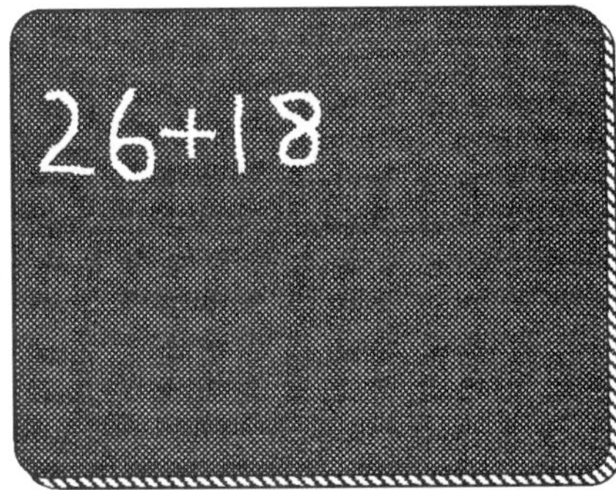

Mrs. Patwin knows that some children are ready to do the problem without using manipulatives to help them, and she wants them to think the problem through first. Sometimes she has them share their ideas and sometimes she doesn't. While she wants to challenge them, at the same time she doesn't want those children who still need the models to feel that there is something wrong with their approach. This time she asks everybody to use their cubes before beginning any discussion. "Now, everybody use cubes to show how you got your answer."

After the children have worked out their answers with the cubes, she asks, "Who would like to share their answer?"

Several hands go up, and the teacher calls on Samira, who says, "I got 44."

Mrs. Patwin writes 44 on the board.

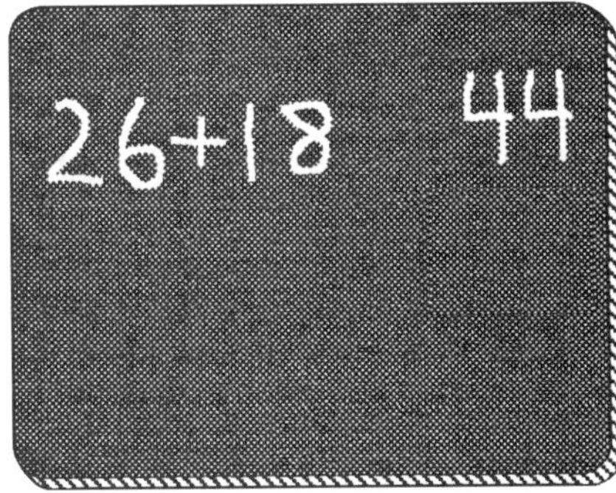

"Does anybody have a different idea?"

Deanna raises her hand and shares her idea. "I think it's 40."

Mrs. Patwin writes Deanna's answer on the board.

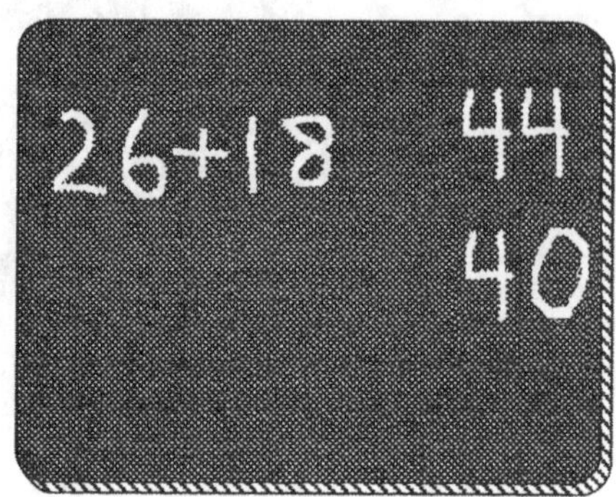

Even though Deanna is the only one to raise her hand to share a different idea, she is not afraid to tell what she got. That is because Mrs. Patwin does not react to any of the answers the children present. The class simply works out the problem and sees what happens. Then, when the children discuss how they solve the problem, the right answer becomes obvious and is handled matter of factly.

Since no one has another answer, Mrs. Patwin asks, "Who would like to share how you got your answer?"

Peter raises his hand. "I put 2 tens and 6 ones on my board, and then I put 1 ten and 8 ones on my board. That made 3 tens and 14, and I know that made 44."

Deanna raises her hand. "I was thinking I had 2 tens and one more ten and that was 30. And then I needed two more to go with the 8 to make ten more, and that's 40. But I forgot I still had some left over that I didn't use to make ten. So I want to change my mind about my answer. Now I think it's 44, too."

Soo Li raises her hand. "I knew when we were thinking about it that 26 and 20 make 46. But that's two extra. So I could take two off. So I put 2 tens on my board and took two off and that did make 44."

Natasha raises her hand. "I put 26 on my board and then I put 18 on my board and I counted all of them and that made 44."

Mrs. Patwin sees that the class is ready to go on. "Now I am going to put another problem on the board. First use cubes to show this number." She writes 23 on the board and watches as the children put out two tens and three.

"I want you to think about this problem. Thumbs up when you think you know." The teacher finishes writing a subtraction problem on the board.

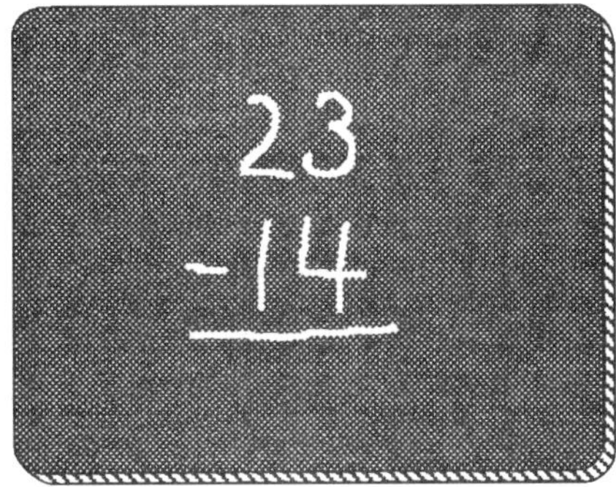

Mrs. Patwin is a bit surprised to see two thumbs put up almost immediately. She waits a few minutes to give more time to the children who need it. Most of the children are using the connecting cubes as they find it hard to do subtraction without manipulatives at this stage. Finally the teacher asks for volunteers to share their answers. She records as the children share their answers. She usually gets more different answers when the children are working on subtraction, as it is more difficult for them than addition. Her children are happy to risk sharing their ideas, including 11, 37, and 9. Mrs. Patwin then asks the children to tell how they got their answers.

Philip starts to explain his answer of 37 when he realizes that instead of subtracting, he added.

Jórge explains, "Well, I thought about 23 take away 10 is 13. But I have to take away four more. So I counted backwards to get nine left."

"Did anybody do it a different way?"

Sara volunteers, "I was thinking 20 take away 10 is 10. Then I can take off four more from that 10 and there is six left. But there are three more, so I knew that six and three made nine."

"So what have we decided 23 − 14 is?" Mrs. Patwin asks. The children agree that the answer is 9.

Mrs. Patwin wants to keep this time short, so she has her helpers gather up the materials so that the class can have some time to work at independent stations. She knows that all the children are making progress, and she is quite impressed with the methods they are using to figure out the answers. She will continue to give the children practice with both addition and subtraction and will add more challenges as she sees that the children are ready for them. She notices that a few children still count on their fingers, even when figuring out answers to ten. Mrs. Patwin will pull these children aside to work with her in a small group before joining the others at the independent stations. She will have them work with smaller numbers on an ongoing basis until they can figure out problems without counting one by one.

About the Activities in Section C

Adding and Subtracting: Shared Experiences

For the teacher-directed activities in Section C, simply offer many whole-class and small-group experiences with addition and subtraction of two-digit numbers. You should repeat these tasks many times, varying the steps and the problems as described. These shared experiences give children the opportunity to benefit from the methods and approaches used by others.

Adding and Subtracting: Independent Practice

Along with the shared experiences, the children will also need lots of independent practice with both addition and subtraction problems. Encourage them to continue using manipulative materials even as they work alone or with partners. The independent activities should also be repeated many times.

Teacher-Directed Activities

1–45 Addition and Subtraction of Two-Digit Numbers .... Whole-Class Activity

Materials: Connecting cubes or beans and cups ▪ Place-value boards (1 per child)

In presenting this approach, it is important to teach the processes of addition and subtraction together right from the beginning so that children learn to always consider the situation and/or check the addition or subtraction symbol to see what to do.

Addition

Have the children create sets of ten, either by making connecting-cube trains or by filling cups with 10 beans each.

It's important to have the sets of tens available so that the problem-solving process is not slowed down. When children see the value of grouping quantities into tens, they are more apt to use the sets of ten to add and subtract instead of counting one by one.

Tell the children that you are going to write a number on the chalkboard and that you want them to use counters to show that number on their place-value boards. Then, write 35.

What's the fastest way to make this number?

Some children will want to count out the 35 counters one by one, putting them into a pile. While it would not be incorrect to do this, it would not lead to the kind of thinking we are trying to support. To encourage children to show 35 with tens and ones, ask them to show it the *fastest* way. The fastest way to build 35 is with 3 tens and 5 loose counters. Practice with this concept is provided by the Chapter 1, Section B activity, Build It Fast (1–29).

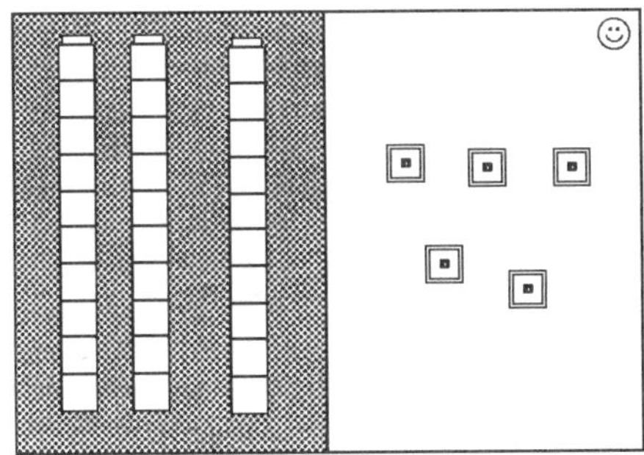

Now I am going to write another number.

Do we need to add or subtract?

We need to add.

Before you use any counters, predict what you think the answer is going to be.

Asking the children to predict what the answer will be serves two purposes: Firstly, those children who can add without the use of models are encouraged to use their best thinking to solve the problem. Secondly, those who still need models are given the opportunity to consider what a *reasonable* answer might be before they actually use the materials.

The next steps should vary.

1. Sometimes, have *everyone* use counters to show their answers even if they were able to figure out the answers without them.

   It is important, especially in the beginning, to validate the use of the models to get answers. Most of the children will need models for quite some time until they become able to think about the relationships without them.

2. Other times, have those children who are willing to share tell what they think the answer will be before they use the counters. Have some of them also share how they got their answers. After each way has been shared, have everyone "prove it" with counters.

   We want those children who see the relationships and who don't need models to be able to use this level of thinking and to share their thinking with others. If everyone is required to show the work with the models, no one is made to feel inadequate because of needing to use models. Also, those who think they know but don't will have to look again when they use the models.

3. Still other times, tell the children to figure out the problem in any way they can. They may choose to use counters or to not use counters. When they seem ready, ask for volunteers to share how they got their answers.

Children should be encouraged to use whatever method works best for them. One way of determining whether or not children feel okay using their own methods is to see if a variety of ways are willingly shared and accepted. If so, you may hear one child say:

> **I knew that 3 tens and 2 tens made 50. Then I figured out that five and seven made 12. Then I knew 50 and 12 is 62.**

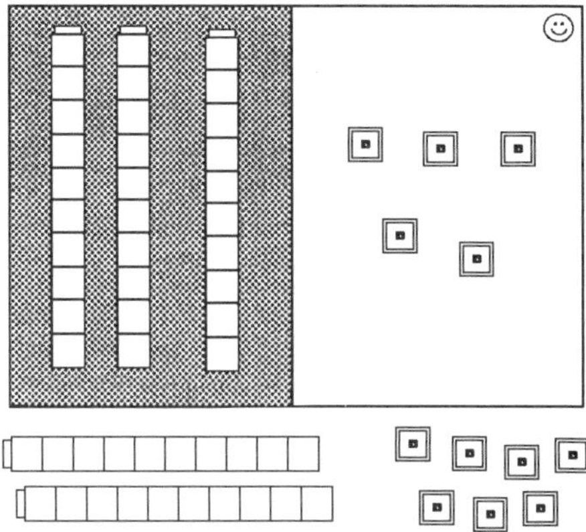

Another child might say:

> **I knew that 30 and 20 was 50. Then I took five off the seven and put it with the other five. That made 10 more and that made 60. I still had two more so that made 62.**

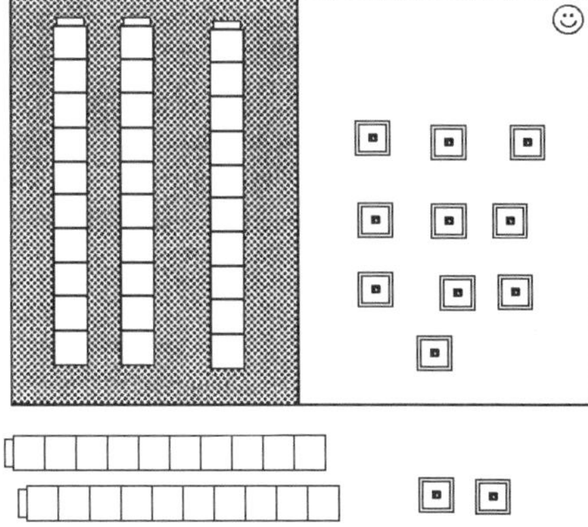

Still other children might say:

> **I got 35 cubes and 27 cubes and I counted all of them and it was 62.**
>
> **I knew 35 and 20 was 55. Then I counted on my fingers seven more, and that was 62.**

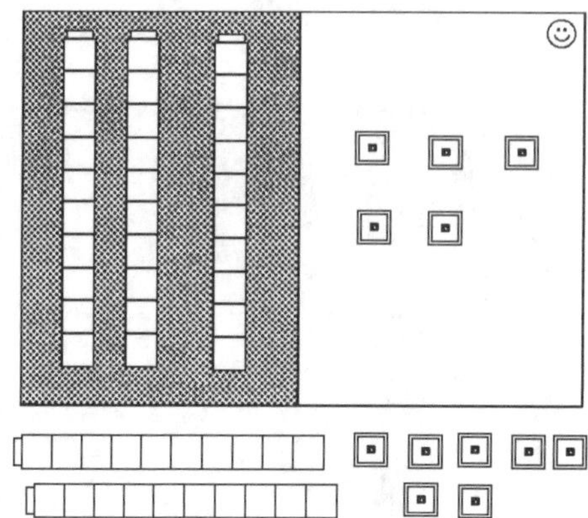

4. However you pose the problems, occasionally ask the children to prove their answers with counters.

The most natural and almost universal way for children to work with these numbers is to start with the tens first and then do what is necessary to add on the rest of the ones.

The children who will have the most trouble figuring out how to think about these problems are those who have already been taught the traditional steps for adding from right to left. This traditional procedure actually interferes with children's ability to sort out their own thinking. However, given enough experiences and the idea that many different ways work, some of these children will become able to use a variety of approaches.

Accept each child's way of getting to the correct answer, but make sure that all the children can also "prove" their answers with models.

Subtraction

I am going to write a number on the chalkboard, and I want you to show that number on your place-value board in the quickest way.

Write 31.

Just as they did before, children should show the number using tens and ones, rather than counting out thirty-one loose counters.

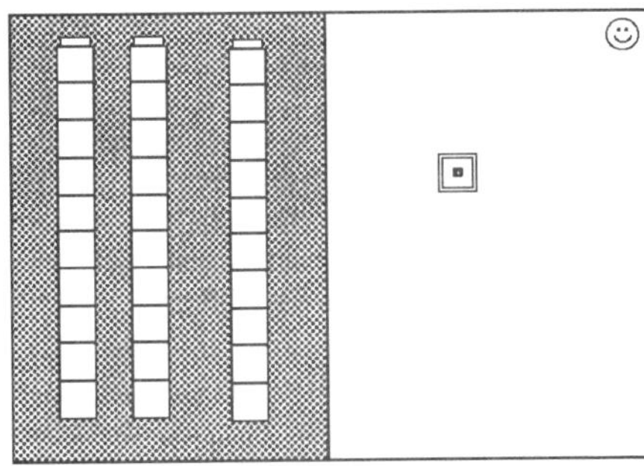

Now I am going to write another number.

Are we going to add or subtract?

We need to subtract.

Think first about what you think the answer will be.

Does anybody want to share what they think the answer is?

Subtraction is much more difficult for the children to work with than is addition. You will probably get a wider range of predictions than you did for addition.

Now I want you to find a way to figure out the answer and prove it with the counters.

Because there are many parts of the subtraction process that can confuse children, you will want to make sure that everyone can show how they got their answer by using models.

Who would like to share how they got their answer?

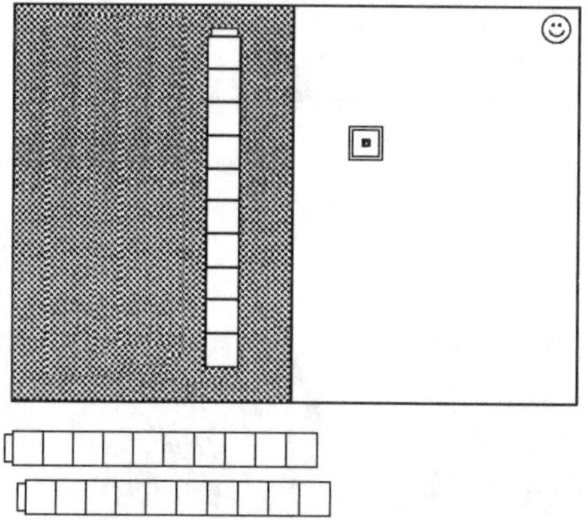

I took 20 away from 31 and that was 11. See, 11.

Then I had to take four more away from the ten.

I had six left and one more and that made seven.

Did anybody do it a different way?

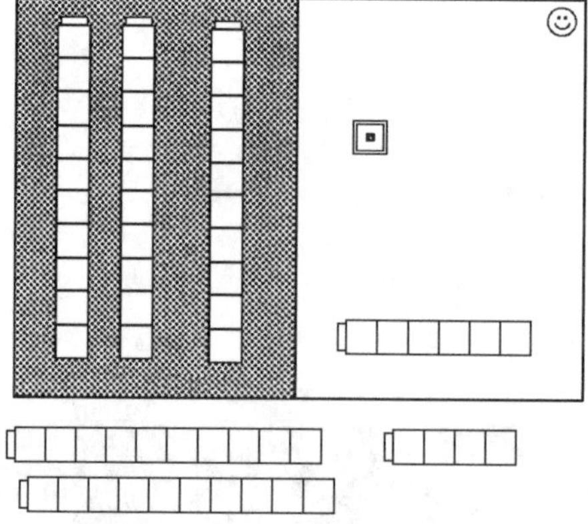

I had 31. Then I counted 24 and took them away from the tens.

I had six and one more. So seven is left.

Children will generally take everything they can away from the tens and then add the leftover ones back at the end. This may be difficult for us to follow since we learned to subtract in a different way, but it makes good sense to the children.

Here are several things to keep in mind as you work with the children on addition and subtraction of two-digit numbers.

1. Present the processes randomly. Sometimes have the children add; sometimes have them subtract. Remind them always to read the signs first to determine what to do if necessary. Sometimes include story problems (as in activity 1–46) and let the children decide whether they need to add or to subtract.

2. Vary the level of difficulty. Include relatively simple problems along with harder ones. Simpler problems will enable those children who are always using the models or counting to get the answers. The simpler problems will allow them to use more sophisticated methods than they can with larger numbers.

   Don't assume that you should introduce addition and subtraction problems that don't require regrouping before introducing problems that do require regrouping. In the long run, doing this makes it harder, and not easier, for children to learn to solve problems. It is misleading to introduce addition and subtraction of two-digit numbers with problems in which no regrouping is necessary. The children have access to the models and will understand the process better if they have to combine numbers to make tens.

3. Don't praise one specific method over another. Help the children clarify their methods, if necessary, by restating what they have said. Sometimes use models to show a method used, but be careful not to pressure the other children to use an idea that makes no sense to them by treating it as being better than other methods.

4. Don't put particular children on the spot to explain their way of getting an answer. Children will be more willing to share if they don't feel pressured to explain what they did.

5. This should be just one of the children's experiences in working independently with place value. Be sure to also work with them in small groups, providing them with the particular level of difficulty most suited to their needs.

1–46 Story Problems

Whole-Class Activity

Materials: Connecting cubes or beans and cups ▪ Place-value boards (1 per child)

Tell a story problem as the children first model it using the cubes and then write the appropriate equation. For example:

John's mother bought a box of oranges. There were 52 oranges in the box. She brought them to school to share with our class. The 29 children each took an orange. How many oranges did John's mother have left?

The children put out 5 sets of ten counters and 2 leftovers, write 52 − 29 and then figure out the answer.

The children then make up their own story problems for others to solve later at the independent station, Solving Story Problems (activity 1–59).

NOTE: In Book Two, *Addition and Subtraction,* counters are used to represent actual objects, people, or animals in story problems. Because objects, people, and animals do not naturally group themselves into tens and ones, the counters should now be considered tools for solving the problems rather than representations of realistic situations.

1–47 Figure It Out

Whole-Class Activity

Materials: None needed.

Present related problems to give children the opportunity to use what they figured out for one problem to help them solve a related problem.

Plus one

$$\begin{array}{r} 24 \\ +\ 1 \\ \hline \end{array} \qquad \begin{array}{r} 36 \\ +\ 1 \\ \hline \end{array} \qquad \begin{array}{r} 43 \\ +\ 1 \\ \hline \end{array} \qquad \begin{array}{r} 98 \\ +\ 1 \\ \hline \end{array}$$

Minus one

$$\begin{array}{r} 23 \\ -\ 1 \\ \hline \end{array} \qquad \begin{array}{r} 22 \\ -\ 1 \\ \hline \end{array} \qquad \begin{array}{r} 19 \\ -\ 1 \\ \hline \end{array} \qquad \begin{array}{r} 64 \\ -\ 1 \\ \hline \end{array}$$

Plus two

$$\begin{array}{r} 35 \\ +\ 2 \\ \hline \end{array} \quad \begin{array}{r} 43 \\ +\ 2 \\ \hline \end{array} \quad \begin{array}{r} 16 \\ +\ 2 \\ \hline \end{array} \quad \begin{array}{r} 38 \\ +\ 2 \\ \hline \end{array}$$

Minus two

$$\begin{array}{r} 35 \\ -\ 2 \\ \hline \end{array} \quad \begin{array}{r} 43 \\ -\ 2 \\ \hline \end{array} \quad \begin{array}{r} 16 \\ -\ 2 \\ \hline \end{array} \quad \begin{array}{r} 38 \\ -\ 2 \\ \hline \end{array}$$

Plus 10

$$\begin{array}{r} 17 \\ +10 \\ \hline \end{array} \quad \begin{array}{r} 32 \\ +10 \\ \hline \end{array} \quad \begin{array}{r} 19 \\ +10 \\ \hline \end{array} \quad \begin{array}{r} 47 \\ +10 \\ \hline \end{array} \quad \begin{array}{r} 42 \\ +10 \\ \hline \end{array}$$

Minus 10

$$\begin{array}{r} 15 \\ -10 \\ \hline \end{array} \quad \begin{array}{r} 25 \\ -10 \\ \hline \end{array} \quad \begin{array}{r} 32 \\ -10 \\ \hline \end{array} \quad \begin{array}{r} 18 \\ -10 \\ \hline \end{array} \quad \begin{array}{r} 63 \\ -10 \\ \hline \end{array}$$

Adding 4 to numbers ending in 3

$$\begin{array}{r} 13 \\ +\ 4 \\ \hline \end{array} \quad \begin{array}{r} 23 \\ +\ 4 \\ \hline \end{array} \quad \begin{array}{r} 43 \\ +\ 4 \\ \hline \end{array} \quad \begin{array}{r} 63 \\ +\ 4 \\ \hline \end{array}$$

Adding 5 to numbers ending in 9

$$\begin{array}{r} 9 \\ +\ 5 \\ \hline \end{array} \quad \begin{array}{r} 29 \\ +\ 5 \\ \hline \end{array} \quad \begin{array}{r} 49 \\ +\ 5 \\ \hline \end{array} \quad \begin{array}{r} 69 \\ +\ 5 \\ \hline \end{array}$$

Plus 20

$$\begin{array}{r} 34 \\ +20 \\ \hline \end{array} \quad \begin{array}{r} 43 \\ +20 \\ \hline \end{array} \quad \begin{array}{r} 67 \\ +20 \\ \hline \end{array} \quad \begin{array}{r} 36 \\ +20 \\ \hline \end{array}$$

Minus 20

$$\begin{array}{r} 36 \\ -20 \\ \hline \end{array} \quad \begin{array}{r} 42 \\ -20 \\ \hline \end{array} \quad \begin{array}{r} 56 \\ -20 \\ \hline \end{array} \quad \begin{array}{r} 63 \\ -20 \\ \hline \end{array}$$

Plus 19

$$\begin{array}{r} 20 \\ +19 \\ \hline \end{array} \quad \begin{array}{r} 22 \\ +19 \\ \hline \end{array} \quad \begin{array}{r} 32 \\ +19 \\ \hline \end{array} \quad \begin{array}{r} 41 \\ +19 \\ \hline \end{array}$$

Minus 19

$$\begin{array}{r} 24 \\ -19 \\ \hline \end{array} \quad \begin{array}{r} 36 \\ -19 \\ \hline \end{array} \quad \begin{array}{r} 52 \\ -19 \\ \hline \end{array} \quad \begin{array}{r} 54 \\ -19 \\ \hline \end{array}$$

Independent Activities

1–48 Partner Add-It

Independent Activity for Partners

Materials: Connecting cubes or beans and cups ▪ Place-value boards (1 per child) ▪ Paper and pencil for each child

One partner puts some counters on the place-value board, while the other partner places some counters on the table just below the board. Each child writes the problem that is created in this way.

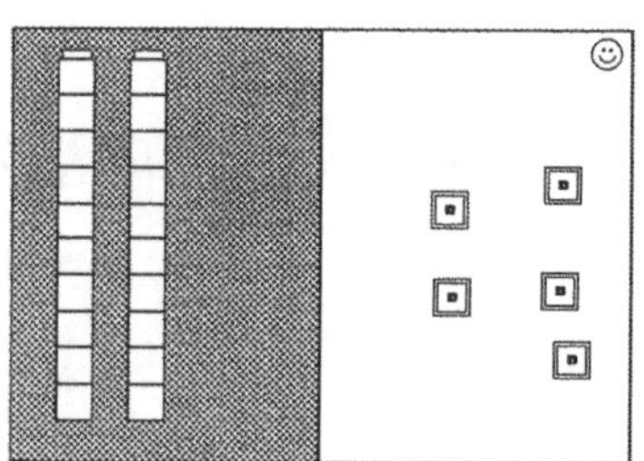

I put 2 tens and five on the board.

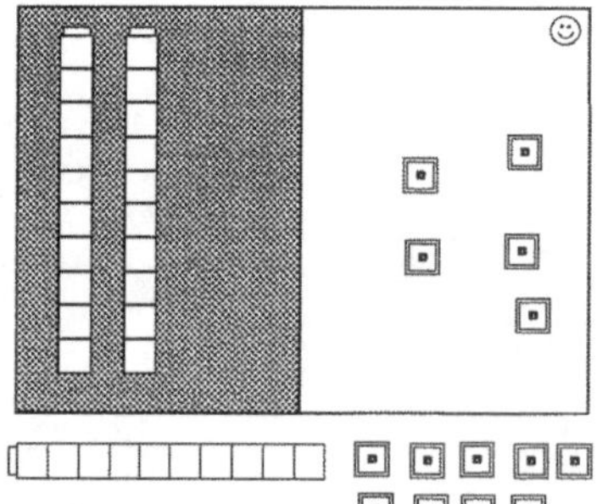

I put 19 under the board.

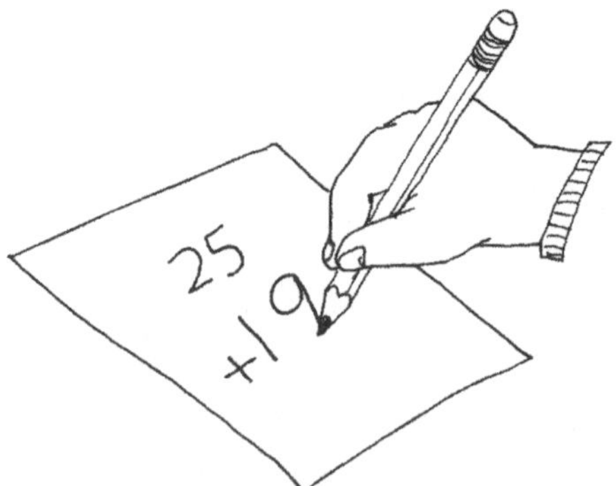

After both partners have written the problem, they each try to figure out the answer without moving the cubes. They then explain their answer to one another by using the cubes.

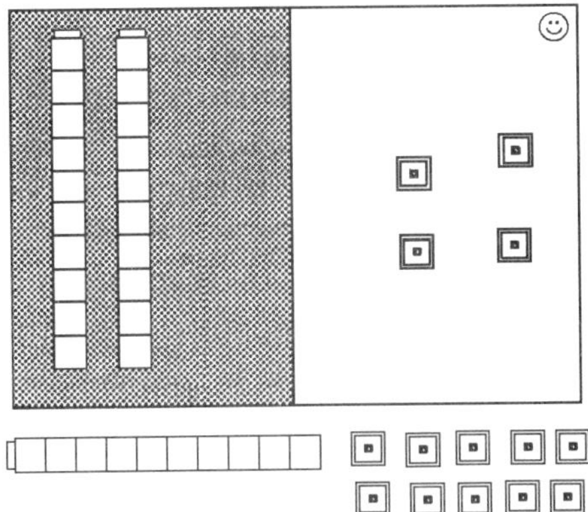

All the tens made 30. I took one cube from the five and put it with the nine and that makes 10, and 30 plus 10 makes 40. Four is left. So it's 44.

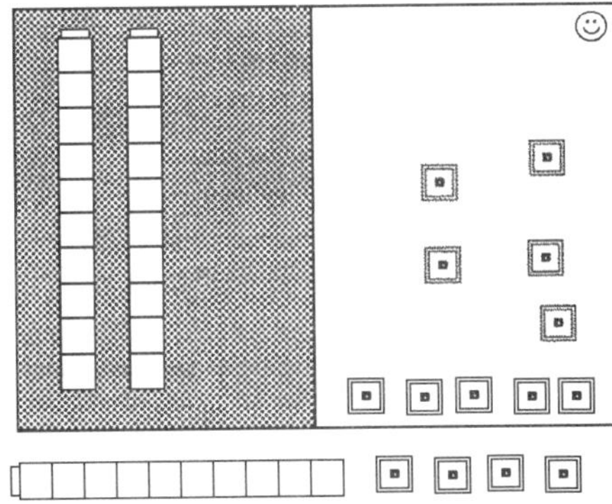

I had 25. I took five off the nine ones and put them with the five on the board. That made 30. Then I added the 14 left to 30 and that was 44.

1–49 Partner Take-Away

Independent Activity for Partners

Materials: Connecting cubes ▪ Place-value boards (1 per child) ▪ Sheet to cover the cubes on the board ▪ Paper and pencil for each child

The first partner uses cubes to show a number on the place-value board. Both partners write that number on their paper. The second partner decides how many cubes to take away from the starting number, and they both write down that amount. The second partner then removes that number of cubes from the place-value board, putting them below the board and covering the amount left. Together the partners figure out how many cubes are under the paper. For example:

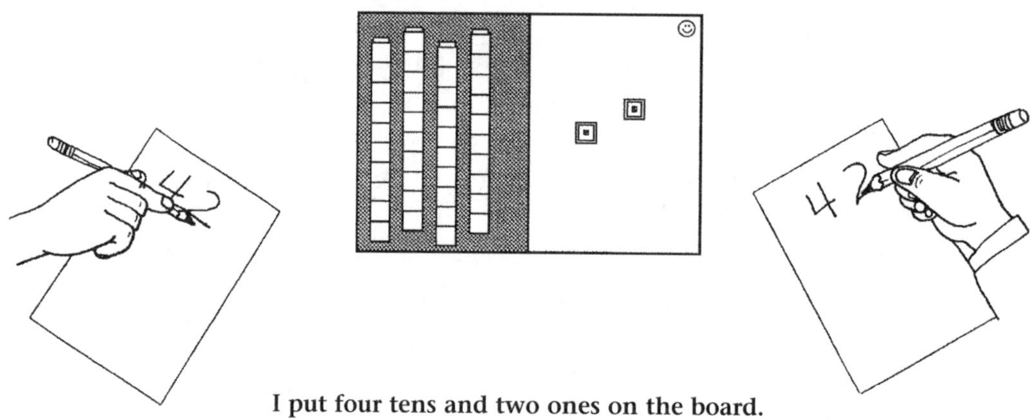

I put four tens and two ones on the board.

Both children write 42 on their paper.

The second partner says, "I'm going to take away 15."

The children each write −15.

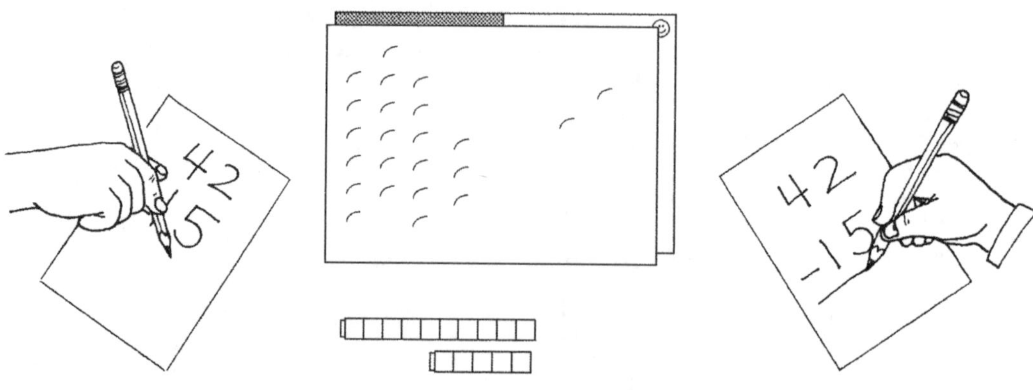

The second partner takes 15 cubes from the board and covers the remaining cubes. Individually the children figure out how many cubes are under the paper. They write their answers. They then lift the covering sheet and check to see if they got the right answer.

For the next problem, the two children trade roles and the second partner builds the starting number. They continue playing, trading roles with each new starting number.

1–50 Roll and Add .... Independent Activity

Materials: Connecting cubes or beans and cups ▪ Place-value boards (1 per child) ▪ Roll-and-Add Worksheet [BLM #133] cut into strips ▪ 0 tens to 5 tens Number Cubes (1 per child) ▪ 4–9 Number Cubes (1 per child)
Preparation: Make a tens number cube by writing "0 Tens, 1 Ten, 2 Tens, 3 Tens, 4 Tens," and "5 Tens," one on each face of a wooden cube. Write the numerals 4 to 9 on a wooden cube to make a 4–9 number cube. Note that copies of the Roll-and-Add Worksheet can be used repeatedly because new problems will be created each time the children roll the number cubes.

By rolling a tens number cube along with a ones number cube, the children create a variety of addition problems. They record their problems on open-ended problem strips, filling as many strips as time allows.

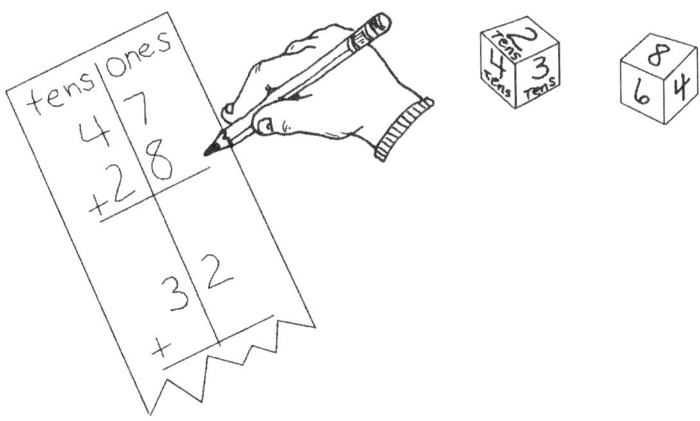

I rolled two tens and eight. I wrote 28.

For each problem, children manipulate cubes on their place-value boards to help them figure out the answer.

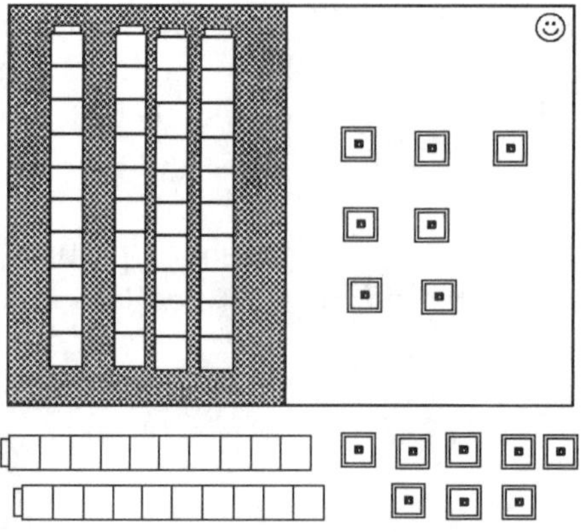

I built 47 and 28.

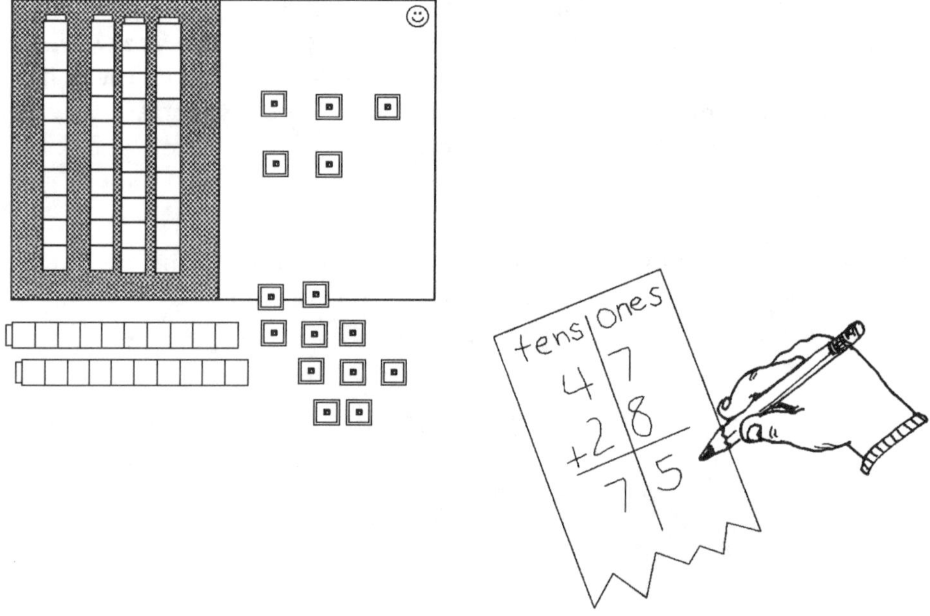

I put two with the eight. That makes a ten.
I have seven tens and five leftovers.

1–51 Roll and Subtract

Independent Activity

Materials: Connecting cubes or beans and cups ▪ Place-value boards (1 per child) ▪ Roll-and-Add Worksheets [BLM #133] cut into strips ▪ 0 tens–5 tens Number Cubes (1 per child) ▪ 4–9 Number cubes (1 per child)

Using a tens and a ones number cube and the Roll-and-Subtract Worksheet strips, the children create a variety of subtraction problems, filling as many strips as time allows.

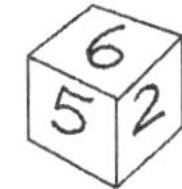

My tens cube has three and my ones cube has six.

I can write 36 under the 86.

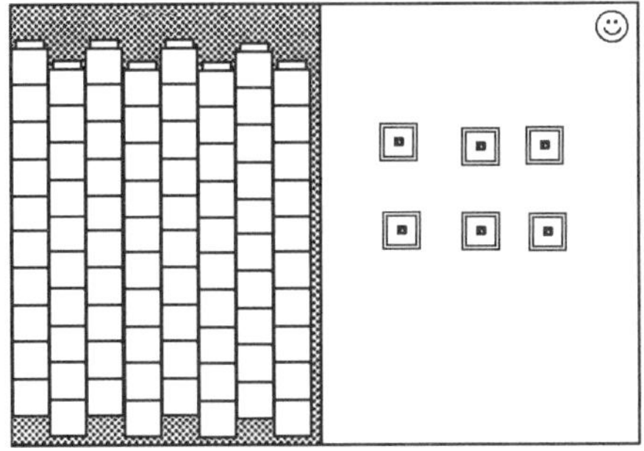

I built 86 on my board.

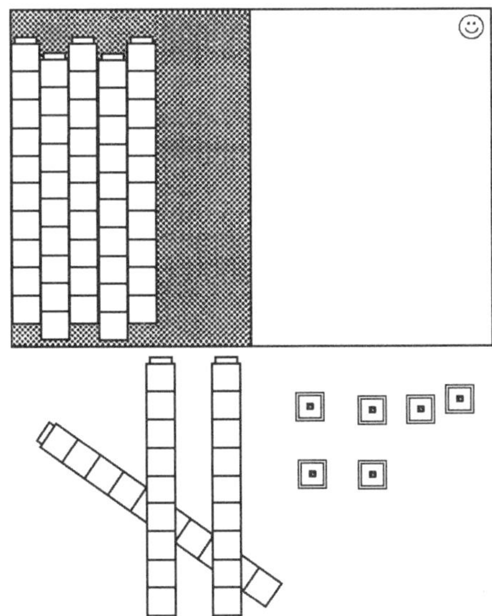

I took 36 away so 50 is left.

1–52 Add 'Em Up: Lots of Lines

Independent or Partner Activity

Materials: Lots-of-Lines task cards (prepared for activity 1–32) • Counters, paper, and pencils

The child selects two Lots-of-Lines task cards, determines the number of counters that can be placed along each line, and then adds the two quantities. For example:

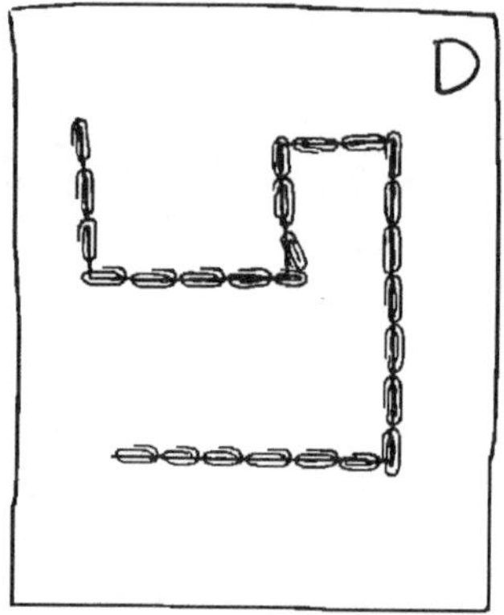

I fit 26 paper clips along this line.

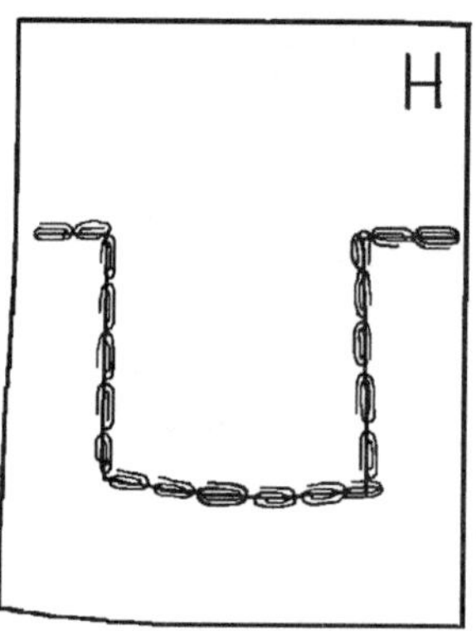

I fit 20 along this line.

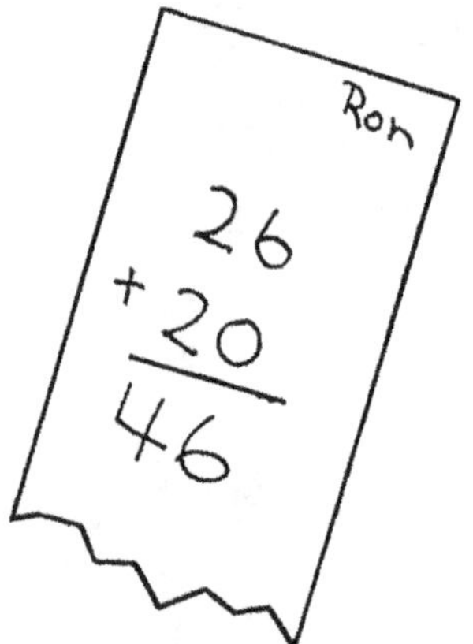

Variation: The children work as partners, each fitting counters along the line on one card. The partners then add the two quantities together.

1–53 Add 'Em Up: Paper Shapes

Independent or Partner Activity

Materials: Paper Shapes task cards (prepared for activity 1–33) ▪ Connecting cubes, wooden cubes, or Color Tiles ▪ Paper and pencils

The child selects two Paper Shapes task cards, determines the number of counters it takes to fill each shape, and then adds the two quantities. For example:

This one holds 30.

This one holds 54.

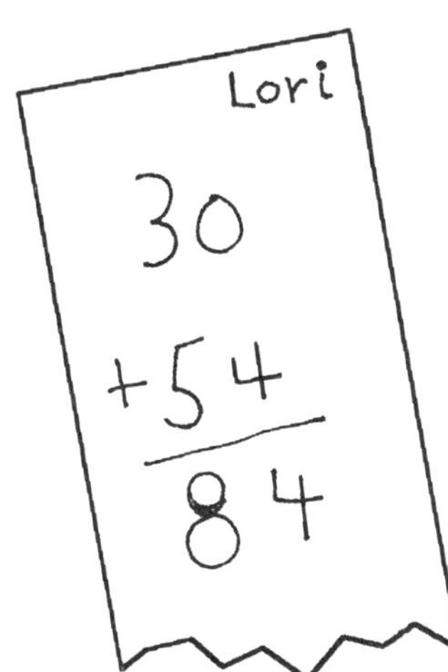

Variation: The children work as partners, each filling one shape. The partners then add the two quantities together.

1–54 Add 'Em Up: Measuring Things in the Room

Independent or Partner Activity

Materials: Measuring Things cards (prepared for activity 1–38) ▪ Counters ▪ Paper and pencils

The child selects two Measuring Things cards and uses counters to measure the length of the actual objects shown on the cards. Then the child adds the two lengths together. For example:

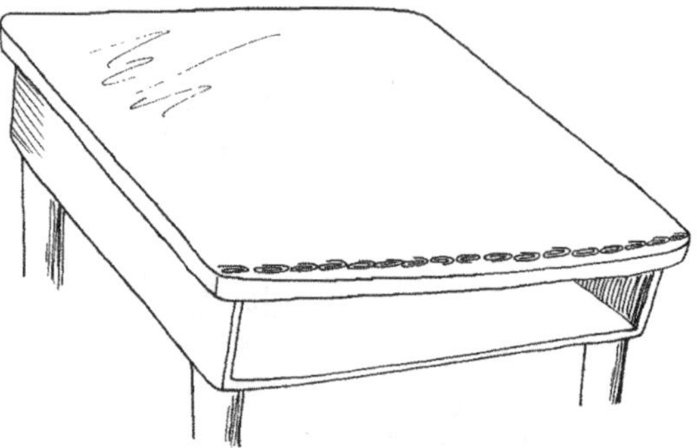

I measured my desk. It was 17 paper clips wide.

I measured a paintbrush. It was 12 paper clips long.

Variation: The children work as partners, each measuring the length of one object in the room. The partners then add the two quantities together.

1–55 Add 'Em Up: Yarn .... Independent or Partner Activity

Materials: Lengths of yarn labeled with letters written on masking tape (prepared for activity 1–34) ▪ Counters ▪ Paper and pencils

The child uses the counters to determine the length of two pieces of yarn and then adds the two quantities together.

First I measured yarn G. It was 19 cubes long

Yarn L was 25.

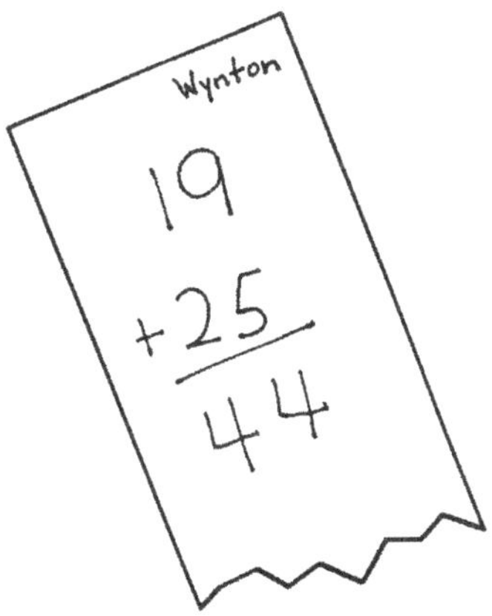

Variation: The children work as partners, each measuring one length of yarn. The partners then add the two quantities together.

1–56 Add 'Em Up: Yarn Shapes

Independent or Partner Activity

Materials: Lengths of yarn labeled with letters written on masking tape (as prepared for activity 1–34) • Connecting cubes, wooden cubes, or Color Tiles • Paper and pencils

The child makes a closed shape with one piece of yarn and determines how many counters it takes to fill that shape. He or she then repeats the process with a second length of yarn and adds the two quantities together. For example:

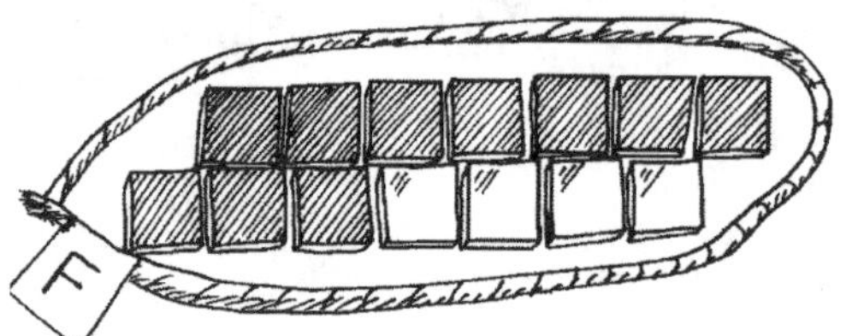

I filled this shape with 14 tiles.

I made this shape. It took 36 tiles.

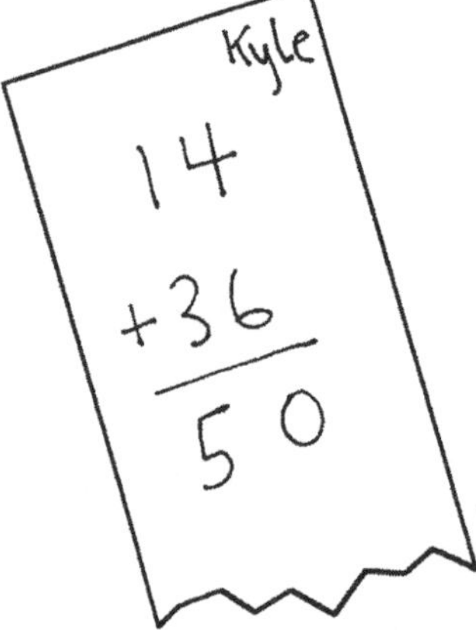

Variation: The children work as partners, each making a different yarn shape and determining how many counters will fill that shape. The partners then add the two quantities together.

1–57 Add 'Em Up: Containers

Independent or Partner Activity

Materials: Various empty containers (margarine tubs, cans, jars, small boxes) labeled with letters (as prepared for activity 1–36) ▪ Counters ▪ Paper and pencils

The child selects two containers, determines the number of counters each will hold, and then adds the two quantities.

Container D held 19 wooden cubes.

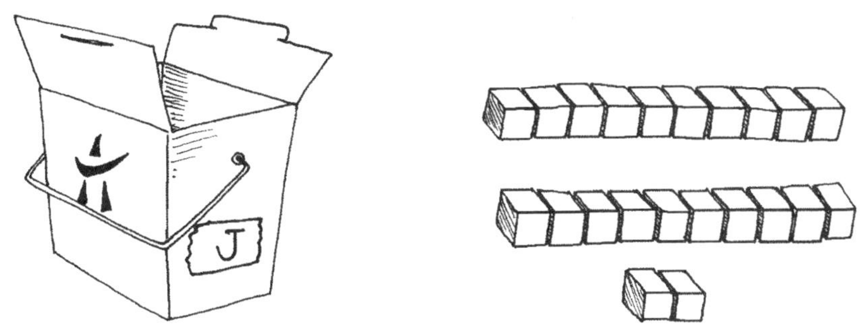

This container held 22 wooden cubes.

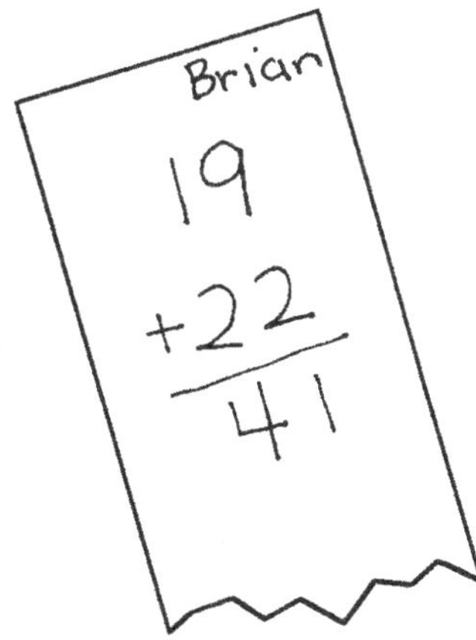

Variation: The children work as partners, each choosing a different container and determining how many counters will fill it. The partners then add the two quantities together.

1–58 Add 'Em Up: Cover It Up

Independent or Partner Activity

Materials: A list of things in the room with a measurable surface area (as prepared for activity 1–37) ▪ Counters ▪ Paper and pencils

The child selects two items from the list of things to "cover up," determines the number of counters needed to cover those objects, and then adds the two quantities.

I covered the book. It took 35 tiles.

I covered an eraser. It only took 16 tiles.

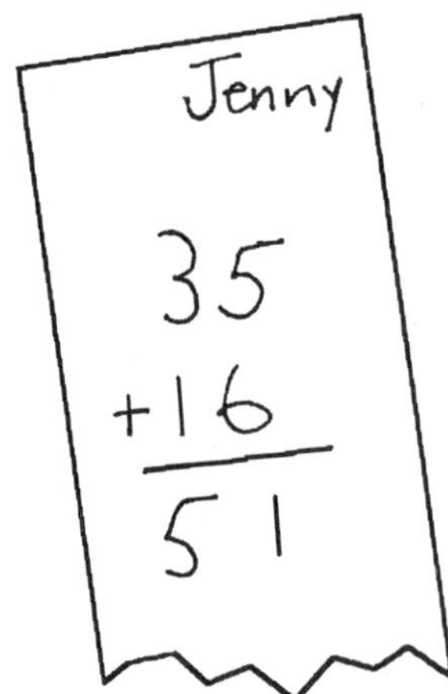

Variation: The children work as partners, each selecting a different item and determining how many counters are needed to cover it. The partners then add the two quantities together.

1–59 Solving Story Problems .... Independent Activity

Materials: Connecting cubes ▪ Place-value boards (1 per child) ▪ Story problems written by you or the children ▪ Paper and pencils

As suggested by the teacher-directed activity Story Problems (1–46), the children write their own story problems. Make a collection of these problems available for independent use by children working either with or without models.

Extension: Real-World Problems

You can provide "real-world" story problems for the children to solve using cubes, as needed, to help them find the totals.

Catalogs After giving the children some instruction in writing dollar-and-cents amounts, have them use a mail-order catalog to choose two or more things that they would like to buy. Have them then figure out how much their order would cost.

Newspapers Have the children look through newspaper advertisements to find two or more things that they would like to buy. Then have them determine the total amount that they would need to spend for these things.

Menus Provide children with restaurant menus. Have them choose things that they would like to order. Then have them add up the cost of the meal.

When textbook or curriculum objectives are:

- Introducing multiplication
  - Multiplying single-digit numbers

Then you want to teach

Beginning Multiplication

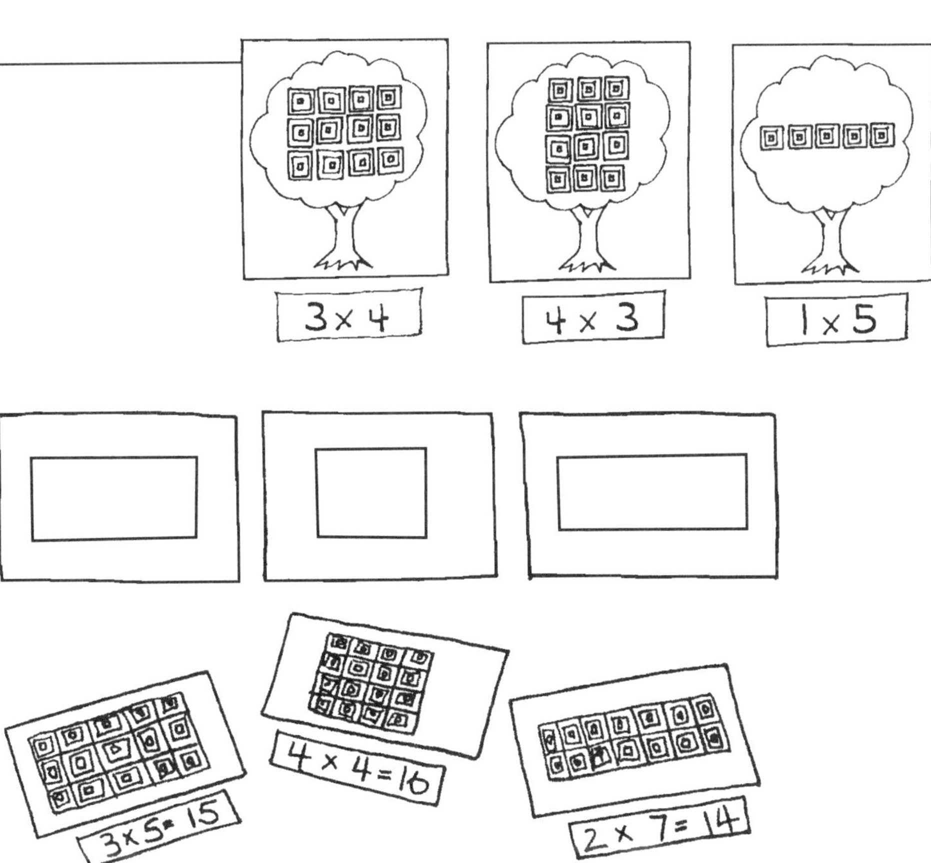

What You Need to Know About Beginning Multiplication

Knowing how to do "times" is for many children an entry into the world of the "big kids." Many children get parents or older brothers and sisters to teach them. Then they come to school, filled with pride, eager to show the teacher what they can do. Many children who can rattle off such phrases as "ten times ten is one hundred" or "two times three is six" have no idea what number relationships they are expressing. They do not understand that the numbers they are saying relate to situations in the real world. This misunderstanding of multiplication became evident to one teacher when she asked a child to use counters to show "four times two" and the child arranged the counters this way:

Children gain an understanding of multiplication through concrete experiences, not through work with symbols.

Although knowing how to recite multiplication equations may satisfy some children's wish to do the kind of math their older brothers or sisters do, this is often of no use to them. Knowing how to multiply is important, but it is useful only if that knowledge is based on real meaning and understanding. Your first goal should be to help children become familiar with the process of multiplication as it appears in the real world. Once children understand the multiplication process as it applies to a variety of situations and can interpret and use the appropriate language, they will be ready to focus on the numerical relationships and patterns that are involved.

Multiplication and addition require different kinds of thinking.

Even though multiplication and addition are related, and multiplication can be thought of as repeated addition, the kind of thinking required for each differs. Truly understanding multiplication requires children to think in terms of groups of things rather than in terms of individual things. We sometimes assume that children think about multiplying groups in the same way that we as adults do, when often they are merely counting or adding. From children's perspective, the groups they are counting just happen to be made up of equal numbers of things. Children often do not really focus on multiple groups. We can prepare them for more formal work with multiplication by giving them lots of opportunities to create, describe, and count equal groups.

Certain language patterns are used to describe multiplication situations. Some are more easily interpreted by young children than others. For example, it is relatively easy for children to consider the phrases "two rows of three tiles" and "four piles of six buttons." However, phrases like "twice as many," "three times as many," or "three per box" are more difficult for them to conceptualize. We should be aware of this when we present such situations to young children.

Children need to learn how to symbolize multiplication situations and to associate the real experiences with the symbols. This may seem obvious to us, but it is not something that comes automatically to children. They often treat multiplication equations as something different from the multiplication situations they describe. When children deal with the symbols and the experiences as two different things, they have trouble using symbols appropriately. One way to help children make the connection between the two is to have them express the symbols in words. For example, they should read $2 \times 4 = 8$ as "two groups of four equals eight." Having them interpret the symbols by describing related situations is another way to help them make this important connection. For example, they might interpret 2×4 by telling a story about two rabbits that each ate four carrots.

We need to help children connect their real experiences to the symbols used to represent those experiences.

Once children understand the process of multiplication and can represent multiplication situations with symbols, they are ready to focus on the number patterns and relationships that will help them internalize the basic multiplication facts. They should spend much of their time exploring and recording multiplication patterns. The search for patterns and relationships will help children learn multiplication facts in a much more powerful way than they would by simply memorizing the times tables.

Exploring patterns and relationships in multiplication aids in the learning of multiplication facts.

Teaching and Learning Beginning Multiplication

The activities in this chapter help children develop an understanding of beginning multiplication as they learn to recognize the process and the language of multiplication. Through the teacher-directed activities, you can help children interpret the language of multiplication, act out multiplication situations, and record their multiplication experiences. The independent activities provide children with the ongoing practice they need in order to become comfortable interpreting multiplication situations and solving multiplication problems.

Using the Chapter

The "Meeting the Needs of Your Children" charts in the introduction to this book and in the *Planning Guide* that accompanies this series offer detailed information that can help you plan how to use the chapter's activities. The following are general suggestions for using the activities with different groups of children.

Kindergarten and First Grade The activities in this chapter are not appropriate for kindergarten or first-grade children.

Second Grade Whether or not children are expected to begin working with multiplication in second grade varies from one school district to another. The activities in this chapter are appropriate for use with second-grade children toward the end of the school year.

Third Grade Learning to multiply is a central endeavor of third-grade students. Helping children build a base for understanding multiplication is vital and so is the focus of the activities in this chapter.

Children with Special Needs If you have special-needs children who need help with multiplication concepts, this chapter can be of help as it deals with the very basic ideas of multiplication. Refer to the "Meeting the Needs of Your Children" charts to find activities appropriate for the children with whom you are working.

Goals for Children's Learning*

Goals

When presented with a variety of multiplication situations, children will be able to:

- Recognize that multiplication is the process of counting equal groups
- Interpret the language of multiplication as it occurs in story problems

* Adapted from *How Do We Know They're Learning? Assessing Math Concepts.*

- See relationships between multiplication problems
- Interpret multiplication situations
- Write multiplication problems to describe appropriate situations
- Solve multiplication problems
- See patterns in multiplication situations

Analyzing and Assessing Children's Needs

When your focus is on helping children develop understanding of beginning multiplication, it is not enough to know whether they have memorized the multiplication tables. It is also not enough to know whether they can complete pages of multiplication problems. Instead, you need to know whether they can recognize and use multiplication to solve problems in a variety of settings. We can determine what children understand about beginning multiplication by observing them at work. Keep in mind that our goal for children working in this chapter is primarily to develop the language of multiplication while becoming familiar with the multiplication process as it occurs in the real world. The following questions can help you look at the stages through which children move as they develop an understanding of multiplication.

Questions to Guide Your Observations*

Questions

Interpreting the Language of Multiplication

- Can the children interpret the language of multiplication by modeling *rows of, groups of, stacks of,* and *piles of* objects?
- Are they flexible in their use of this language?
- Can they easily distinguish, for example, between 3 rows of 4 and 4 rows of 3, or do they have difficulty?
- Do they need to have the problems simplified in order to interpret them?

Interpreting Simple Story Problems

- Can the children interpret multiplication story problems using physical models and/or drawings?
- Do they interpret the problem with ease or with difficulty?
- Do they need any prompts or hints?

Interpreting Symbols

- Can the children interpret multiplication equations using models?
- Can they read the equation using natural language? using formal language?
- Can they make up a story to go with the multiplication situation?

* Adapted from *How Do We Know They're Learning? Assessing Math Concepts.*

Writing and Reading Equations

- Can the children write equations to represent story problems? Is this easy for them or is it challenging?
- Can they read the equations back?
- Do they know how the numbers connect to the situation in each story?
- Do they use natural language or formal language?
- If the child does not know the formal way of recording a multiplication story, can he or she represent the story symbolically in some way?

Solving Multiplication Problems

- When the children solve multiplication problems, what kinds of strategies do they use?

  Do they count all the objects?

  Do they begin counting by groups and then finish counting by ones?

  What groups can they count by? by fives? by tens? by groups of other sizes?

  Do they use relationships to figure out answers? (If so, they might say, for example, "I have four rows of six. I know that two rows of six would be 12, and this is 12 and 12, so the answer is 24.")

  Do they appear to "just know" the multiplication fact?

  Do the children use multiplication appropriately when solving problems or do they still tend to count or add instead?

Seeing Patterns in Multiplication

- Do the children comment on or make use of the multiplication patterns when solving problems?

Meeting the Range of Needs

Children working at many different levels can benefit from the activities in this chapter. The way children determine the answers will vary according to their degree of ease with multiplication.

Some children will count by ones to determine the totals.

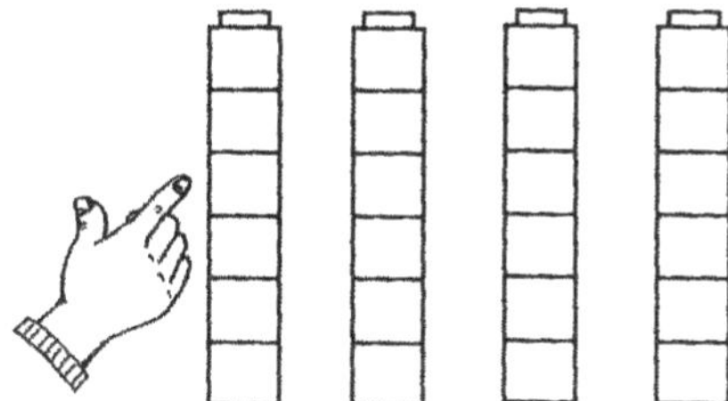

One, two, three, four, ...

Others will count by groups.

Three, six, nine, twelve, fifteen.

Still others will determine the answers they don't know by using multiplication facts that they do know.

I know that two groups of three is six. But there's four groups of three, so that makes six and six more and that makes 12.

If you have some children who are having trouble interpreting the language of multiplication, give them extra support as described in Building Models of Multiplication Problems (activity 2–4).

A Classroom Scene

Mrs. Wezeman, a third-grade teacher, has asked several children to work in a small group while the rest of the class work independently at the multiplication stations. She asks Jennie and Eric to hand out individual chalkboards to the group. She tells them to choose and write a multiplication number pattern while they are waiting for her to start the lesson. Having done this before, they know what to do with little instruction.

Mrs. Wezeman then dismisses the rest of the class a few children at a time to begin their independent-station work. The children have been working with the multiplication stations for several days.

Independent Activity: *Multiplication Stations*

Yesterday the teacher introduced one new activity, Lots of Rectangles (2–23). Today she has her eye on that station to make sure that the children understand what to do. There seems to be some confusion—Tina and Suzanne are arguing over how the task is to be done while Jay watches them doubtfully. The teacher sends Nada to the station to show the children how to do the task, remembering that Nada had no problems working there yesterday. Mrs. Wezeman notices that Jay decides not to take on the new challenge and moves to a different, more familiar, activity.

Glancing around the room one more time, Mrs. Wezeman sees that everyone has settled in. Nada is happily showing Tina and Suzanne how to do Lots of Rectangles. Now, the teacher can turn her attention to the small group.

Small-Group Activity: *Multiplying with Shape Puzzles*

Mrs. Wezeman feels that the children she has asked to work with her today are ready for a challenge. Mrs. Wezeman wants to show them how to use parentheses to record the two parts of a shape puzzle. She begins by putting a shape puzzle on the table in front of the group.

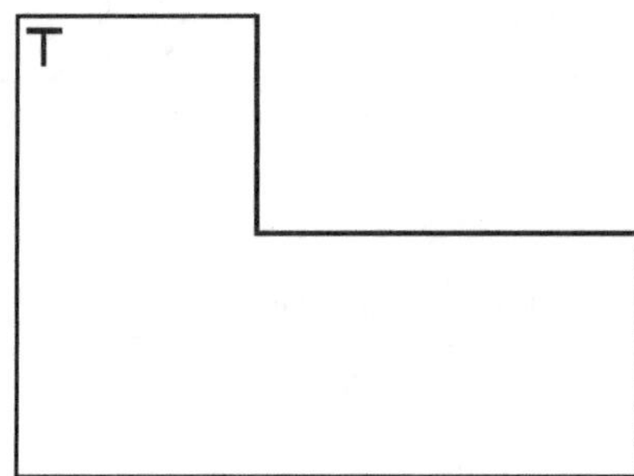

The teacher covers the left side of the puzzle and asks, "Does anybody have an idea of how many rows of Color Tiles we could fit in this part of the puzzle?"

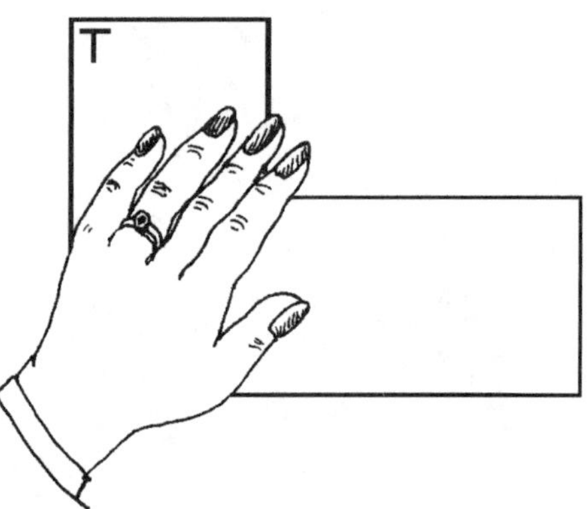

"Maybe four rows of five," says Eric.

"I think six rows of six," Tanisha puts in.

"What about this part?" Mrs. Wezeman says, covering the right side of the puzzle.

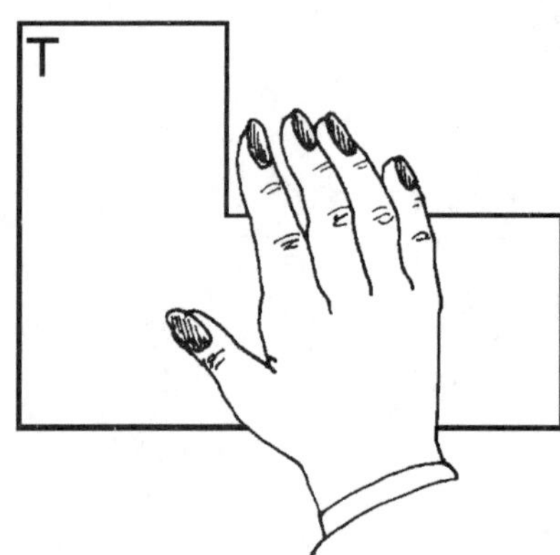

"Looks like three rows of three," says Ty.

"Or maybe four rows of three," says Jennie.

"Let's fill the puzzle with tiles and see if you're right." The children work together to fill the shape. Mrs. Wezeman suggests that they use different colors for each part of the puzzle.

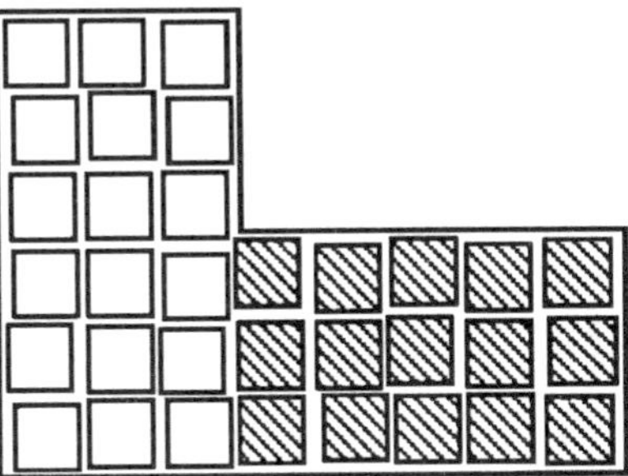

When the shape is filled, Mrs. Wezeman says, "We found yesterday that we could describe Lots of Rectangles with multiplication equations. Do you think we could describe this shape puzzle with a multiplication equation?"

Ty looks doubtful. "That's pretty tricky because the rows aren't all the same."

Mrs. Wezeman smiles. "That's true, but mathematicians have invented a way that we can use to describe the parts when the rows aren't the same."

Tanisha suddenly has an idea. "I bet you do the two parts, just like you did when you covered up one of the pieces."

"Yes, Tanisha, we do the parts separately, and we show what goes together by using parentheses."

Mrs. Wezeman demonstrates how to write "six rows of three" as (6×3) and "three rows of five" as (3×5), and then writes a plus sign between the two expressions to combine them into a single equation.

"Now we'll practice with another puzzle. Try this one. You will need to fill it and then write on your chalkboard what you think will describe it. Use just one color this time so you can decide for yourself what the parts of the shape form."

The children work together to fill the shape and then each writes the equation that he or she thinks describes it.

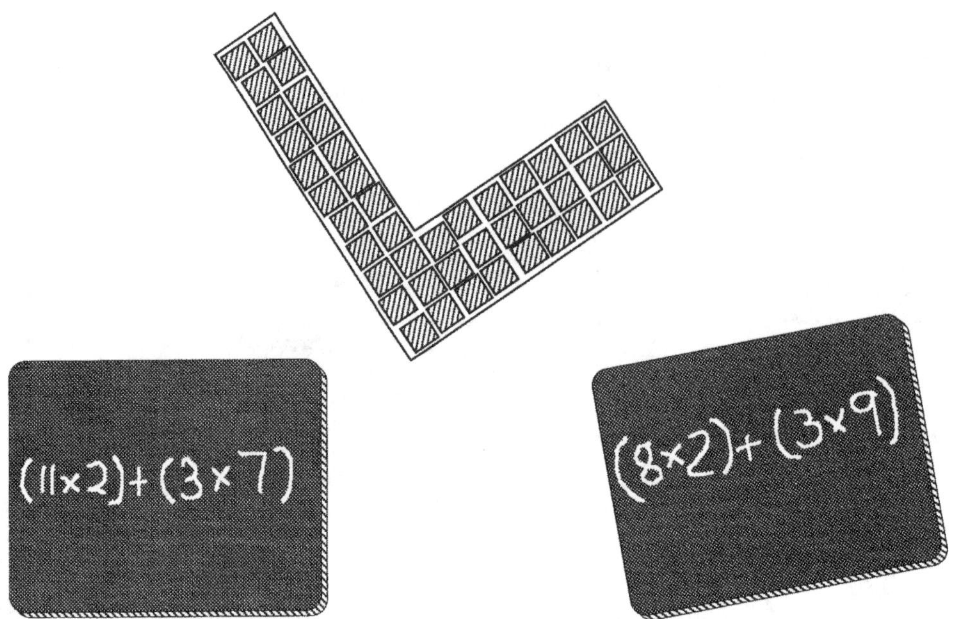

After discussing the new puzzle, Mrs. Wezeman tells the group that they can stay and work with more puzzles on their own or they can choose to work at one of the stations. The group seems intrigued by the new work and all but Jennie decide to stay.

Mrs. Wezeman is now able to move around among the stations, observing and interacting with the children. She goes over to see how Jay is doing with the task he eventually picked. Jay is working with the counting boards and multiplication cards (Counting Boards: Multiplication, activity 2–15). He is placing cubes on the counting boards to represent the multiplication problems on the cards he has chosen.

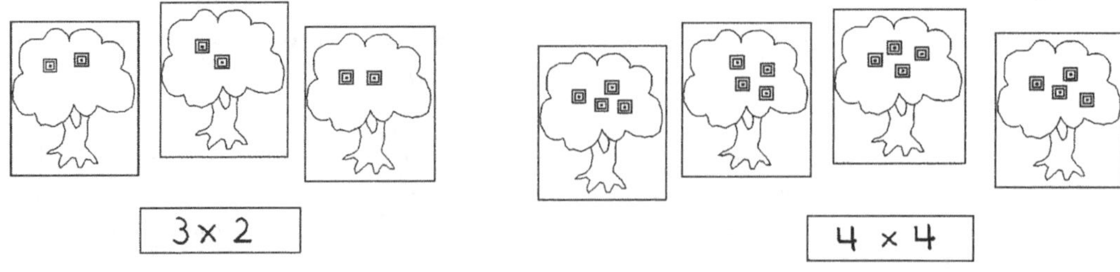

As Mrs. Wezeman watches, she sees that Jay is no longer having trouble interpreting the problems as he had previously. She suggests to Jay that he get some paper and then write the problems and their answers. Jay seems eager to do this. Mrs. Wezeman is quite sure he will be able to do this on his own, but she intends to come back in a few minutes to see whether he is having any trouble.

The teacher moves along to watch the children who are working with Discovering Patterns: Cupfuls (activity 2–18). She always likes to see the different ways in which children arrive at the totals. Since cubes are a little harder to count when they are in cups than when they are in stacks or "towers," Mrs. Wezeman is curious to see whether or not any of the children can figure out how many cubes are in the cups without actually counting each one.

Wyatt has three cups with four cubes in each. He can see the four inside the cups and is counting each one individually.

"Wyatt, do you remember what four and four more is?" asks Mrs. Wezeman.

"That's easy. It's eight," replies Wyatt.

"What can you tell me about the number of cubes in these two cups?"

Wyatt goes back to count the two groups of four. "This is eight, too," he says.

Mrs. Wezeman knows that sometimes children are so intent on learning a new task, they forget to apply what they already know. One of her jobs is to help children make those connections. "Do you think you can figure out how many cubes are in three cups without counting them?" she challenges Wyatt.

"Let's see. Two cups is eight. Three cups is ... twelve," he says, writing the number on his worksheet.

Name Wyatt

0 X 4 = 0
1 X 4 = 4
2 X 4 = 8
3 X 4 = 12
4 X ___ = ___
5 X ___ = ___
6 X ___ = ___
7 X ___ = ___
8 X ___ = ___
9 X ___ = ___

"What about four cups?" asks Mrs. Wezeman.

Wyatt hesitates, then puts out four fingers and counts on: "13, 14, 15, 16. I think 16." He writes down 16. "Could you read the problems on your worksheet to me?" Mrs. Wezeman asks Wyatt.

Wyatt reads each line this way:

"One group of 4 is 4.

Two groups of 4 is 8.

Three groups of 4 is 12.

Four groups of 4 is 16."

"How about just reading the list of the amounts you have altogether for the groups."

Wyatt begins, "4, 8, 12, 16... Hey! That's like a pattern we did before!" The boy looks up on the wall where the patterns are displayed. "That's the pattern for sides of squares: 4, 8, 12. Then four more has to be 16, and I know 16 and 4 more is 20, and 4 more is 24. I think it's the same pattern."

"Are you sure it's going to work the same way?" Mrs. Wezeman asks.

Wyatt hesitates. "I'm not *really* sure," he replies.

"Well," Mrs. Wezeman tells him, "keep working and see what happens."

When the conversation between Wyatt and the teacher started, Isabel looked up and listened for a few minutes. She had been very busy counting each one of the cubes in her cups. The interaction between Wyatt and Mrs. Wezeman didn't help her, so she went back to doing the task in her own way.

On the other hand, after Chana listened for a bit, she proceeded to try counting her cups of five cubes by fives instead of counting each cube one by one. Chana determined her total by counting 5, 10, 15, 20. She wrote down 20, but then went back and counted by ones to see if she ended up with the same number.

Mrs. Wezeman knows that some seemingly simple ideas have complexities that she used to be unaware of. She has learned much about how young children think about multiplication by watching them closely while they work. She is pleased with the fact that she has a variety of ways for her class to learn multiplication and is able to provide appropriate practice for all her students. She will continue to give children the opportunity to work with these activities for many more days.

About the Activities

Developing the Concept of Multiplication

Children should begin their work with multiplication by discovering how multiplication situations occur in the real world. The teacher-directed activities in this chapter offer a sequential guide to this discovery. The first thing children need to understand is that multiplication is the process of counting objects by equal groups. This can pose a challenge to some children because it requires them to think about and count *groups* of objects rather than *single* objects. Your role is to help children recognize equal groups and to develop the language of multiplication. Through the teacher-directed activities you will show the children how multiplication can be represented by models of equal groups that can appear as rows, stacks, and piles. Repeat the introductory activities many times so that children can become flexible in their understanding of any description of multiplication. To support the early work with multiplication concepts, you will have the children go on to practice creating and counting groups through independent activities (2–11 to 2–14) that do not require them to write multiplication equations.

Working with Symbols

Too often children learn multiplication facts purely by rote, with no understanding of what they are memorizing. A series of teacher-directed activities, starting with Modeling the Recording of Multiplication Experiences (activity 2–6), helps children make the connection between the symbols and what they stand for.

As soon as the children can interpret a variety of language patterns representing multiplication, begin to demonstrate the way these experiences can be written down. It is critical that children make the connection between the multiplication experiences that they have been having and the symbols used to record them. To help children make that connection, begin recording the experiences with words rather than with multiplication signs, writing, for example, "4 rows of 3 chairs," and "3 stacks of 3 books." Be very careful to use natural language to describe the situations (and not to use the word *times*).

Have the children also use natural language to record until you feel that they can do this with ease. At that point, it will be appropriate to show them a "shortcut" for writing these phrases. Introduce them to the multiplication sign, but continue to use terms such as *rows of,* and *groups of.* Don't begin using the word *times,* as this seems to lead children to disassociate the situation from the symbols.

After the children have had directed practice with writing equations, they can work independently on activities (2–15 to 2–21 and 2–23) that provide them with the opportunity to write multiplication equations in a variety of situations.

Teacher-Directed Activities

2–1 Looking for Equal Groups in the Real World

Whole-Class Activity

Materials: Chart paper for making lists (optional)

To be successful with multiplication, children must understand the idea that multiplication is the process of counting groups that are equal in number. Help the children to find equal groups by suggesting that they look around in the classroom or on walks around the school. Make a list of groups as the children find them. For example:

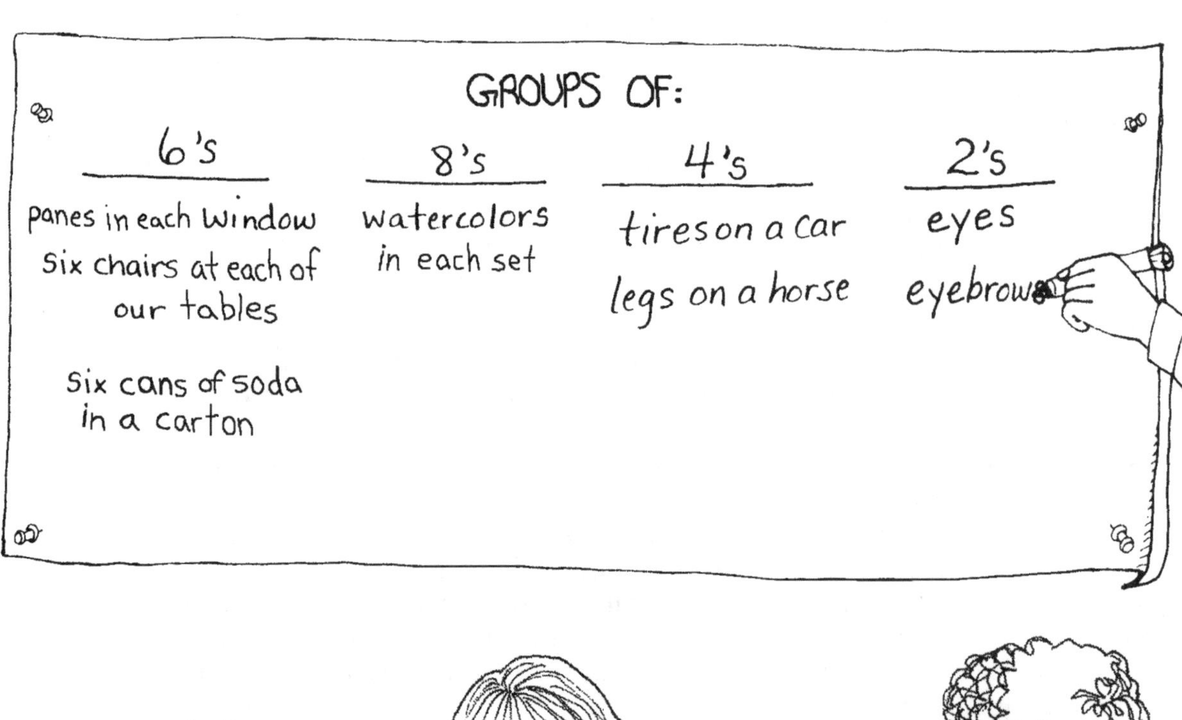

GROUPS OF:
6's
panes in each window
six chairs at each of our tables
six cans of soda in a carton
8's
watercolors in each set
4's
tires on a car
legs on a horse
2's
eyes
eyebrows

Inside the Classroom

I see six panes in the window by my desk. Are there any other windows in our room that have six panes?

I see two more.

Yes, we have three windows, each with six panes. Six panes and six panes and six panes. Today we want to look for other things that have the same number of things over and over, like the windows.

Look at our boxes of watercolors. Do all these boxes each have the same number of colors in them?

Yes. There's eight and eight and eight and eight and eight and eight.

Yes, each box has eight colors. How many eights are there?

Six.

Some of the tables in here have six chairs at them. How many tables have six chairs?

There are four tables that have six chairs.

Yes, we have four tables with six chairs each. Six and six and six and six.

Outside the Classroom

Let's go for a walk and see what kinds of equal groups we can find.

Here are three cars parked together. Look carefully and see if you can find anything on these cars that is the same number as on the other cars.

I see lights. This car has two lights and this car has four lights. That's two and four lights.

We are not looking just for things that look the same. We are looking for things that are the same in number. Are two and four the same in number?

No, but I see something that's the same number. The wheels are the same. See, four and four and four.

Yes, that's right. That's three groups of four.

2–2 Acting Out Multiplication Stories: Using Real Objects

Whole-Class Activity

Materials: Groups of classroom objects

Tell the children multiplication story problems that involve groups of classroom objects and have them act out the stories. For example:

Travis and Kelly line up some chairs. They make three rows with four chairs in each row. How many chairs do they line up?

Dennis, Francis, Lakeisha, Bernadette, and Jamie put five chairs at three tables. How many chairs do they use?

Manuel gives five children two pencils each. How many pencils does he pass out?

Shantea makes four stacks of books. She puts three books in each stack. How many books does she use?

Charlie puts five boxes of crayons on the table. Each box holds eight crayons. How many crayons are there?

Lee puts three erasers on each table in the room. There are six tables. How many erasers does he put out?

2–3 Acting Out Multiplication Stories: Using Counters .... Whole-Class Activity

Materials: Counters ▪ Blank paper (1 sheet per child) or 9" × 12" construction paper (1 sheet per child)

Tell multiplication story problems and have the children act them out using counters to represent groups of people, animals, or objects. When the children listen to these stories, they must be able to determine which objects to model. For example:

There are four houses on Letitia's street. The family in each house has two cars. How many cars is that in all?

Children should be helped to understand that you are asking about the number of *cars* and not the number of *houses*. The number of houses is the clue to the number of groups. This will be confusing for some children, so talk through the problem with them if needed. For example:

Use your counters to show only the cars. How many cars are in front of the first house?

How many cars are in the front of the second house? Show this with counters.

Have you shown the cars for all the houses? Show me the cars in front of the next house.

You made four groups of two cars. How many cars is that in all?

NOTE: If some children need to see the "houses," they might use a counter to represent each house, then put two counters of a different color in front of each "house." When you ask "How many cars?" make sure they count only the counter representing cars and not those representing the houses.

Here are some additional story problems:

- ***Tim had three dogs. He gave each dog two bones. How many bones did he give all his dogs?***
- ***Five girls went to the library. They each checked out three books. How many books did they check out altogether?***
- ***There are five children in Dale's family. Each child gets to carve one pumpkin for Halloween. How many jack-o'-lanterns will they have?***
- ***Robin's mother went shopping for school clothes for her three children. She bought three new shirts for each child. How many shirts did she buy?***

2-4 Building Models of Multiplication Problems

Whole-Class or Small-Group Activity

Materials: Connecting cubes, wooden cubes, Color Tiles, or collections

As children work with multiplication, they must think in terms of kinds of groups as well as the numbers and sizes of groups. They need to have experiences modeling multiplication in many different ways—in towers or stacks, in rows, and in piles or groups.

These activities require the children to interpret various language patterns in order to build the appropriate models. Some children will need to repeat this work over and over again. Build the models along with the children while they are learning.

Towers or Stacks

Alternately use the words *towers* and *stacks*.

Build three towers of five.

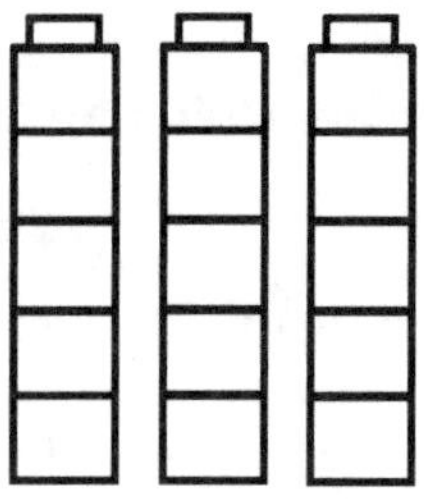

Build two towers of four.

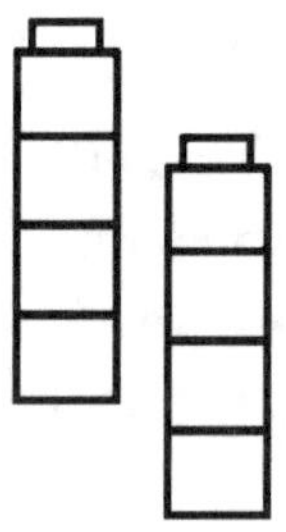

Build four stacks of six.

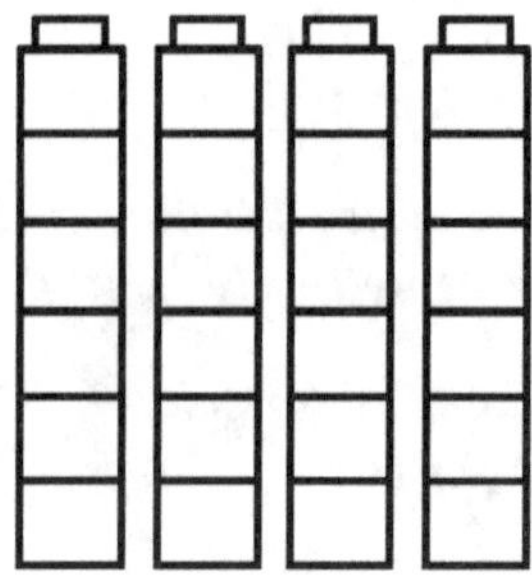

Rows

Define a *row* as being a straight line that goes in the direction that your arms are in when you hold them out to your sides.

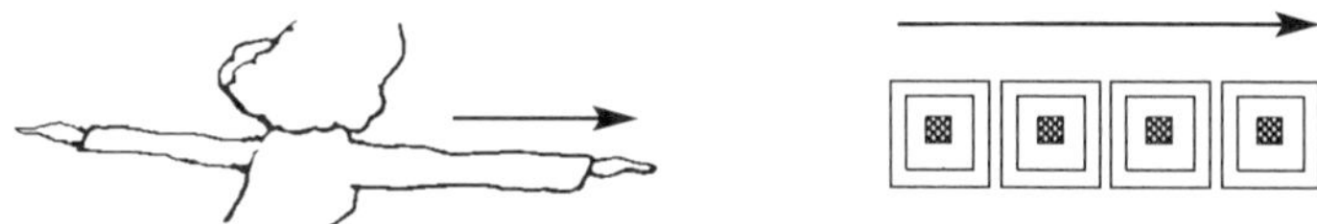

Ask the children to make the rows in their models touch. (This will help prepare them for later work with forming arrays to represent area.)

Make three rows of four.

Make two rows of three.

Groups or Piles

Alternately use the words *groups of* and *piles of*.

Make two piles of five.

Make three groups of four.

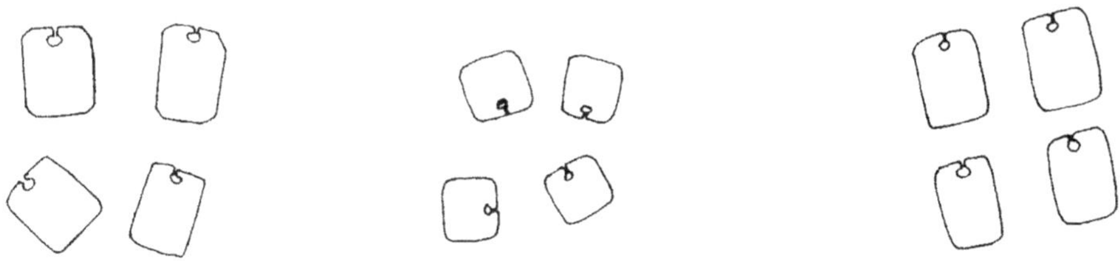

When Help Is Needed If the children are having any difficulty interpreting the language of multiplication, the following procedure can help them remember which number tells *how many groups* and which number refers to *the number in each group*.

Start with one row and add additional rows one at a time. For example:

This is one row of five. Now you build one row of five. How many rows have you built?

One.

Yes, you built one row of five.

Here is another row of five. Now I have two rows of five. You build another row of five. How many rows of five do you have now?

Two rows of five.

As soon as the children are confident, change the number of counters in each row so that they can describe them in a variety of ways. For example, display the following model:

Now what do I have?

Two rows of three.

Yes. Now build four rows of three.

Continue to vary the directions for the children to follow, keeping "rows of three" consistent. Include, for example, "three rows of three" and "seven rows of three."

When the children become successful at following your instructions, give them a variety of directions, sometimes changing the number of rows and sometimes changing the number of cubes in each row. If children appear confused or hesitant, ask them to build one row again and help them clarify their thinking.

2–5 Building Related Models

Whole-Class or Small-Group Activity

Materials: Counters

Once the children are able to build rows, stacks, or groups easily, present them with building tasks based on related multiplication facts, such as *six groups of two* and *two groups of six*. For example:

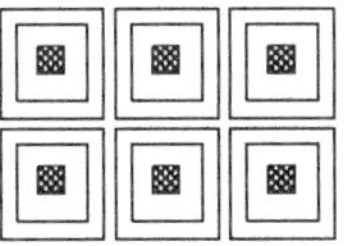

Build two rows of three. How many altogether?

Build three rows of two. How many altogether?

Build three piles of five. How many?

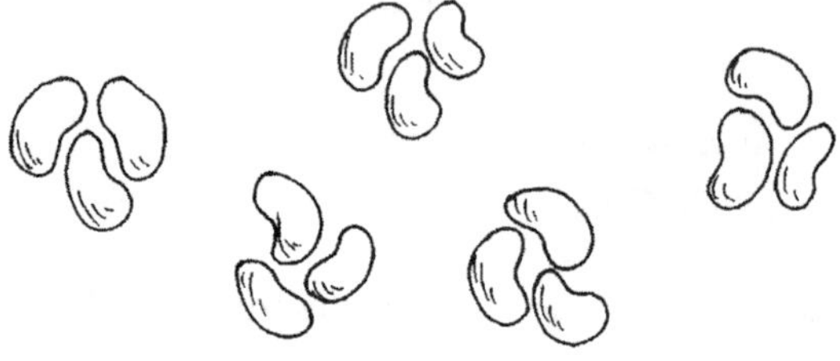

Build five piles of three. How many?

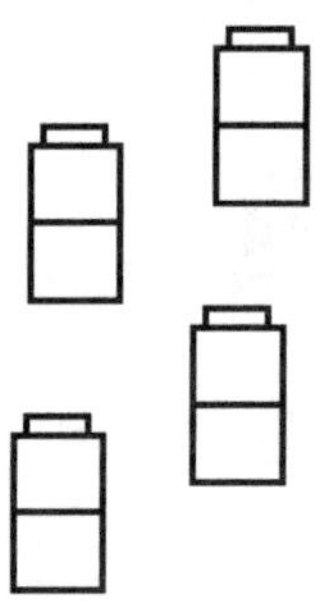

Build four stacks of two. How many?

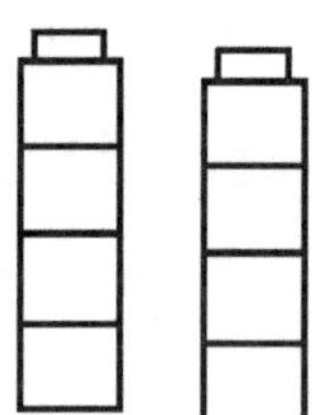

Build two stacks of four. How many?

NOTE: It will not be obvious to some children that the related models require the same total number of cubes. Do not take away the children's opportunity to discover this for themselves. Your responsibility is only to provide situations in which these relationships occur. Children should be allowed the time to discover the relationships on their own. Simply observe the way they solve the problems at this stage of their development.

2–6 Modeling the Recording of Multiplication Experiences

Whole-Class or Small-Group Activity

Materials: Counters or classroom objects that can be counted

When the children can interpret a variety of language patterns representing multiplication experiences, begin to demonstrate how these can be written down. It is critical that the children learn to connect the written symbols with the multiplication experiences they have been having. To help them make that connection, begin recording the experiences using words rather than the multiplication sign. Be very careful not to use the word *times,* as focusing children's attention on the symbol may interfere with their ability to think about the situation.

Tell a multiplication story problem that the children can act out either with actual classroom objects or with counters.

Misty stacks books into two piles. She puts four in each pile.

As the children are acting out the problem, say and write the following:

How many stacks is Misty making?

Two.

Write: 2 stacks of

How many books are in each pile?

Four.

Write: 2 stacks of 4

How many books altogether?

Eight.

Write: 2 stacks of 4 = 8

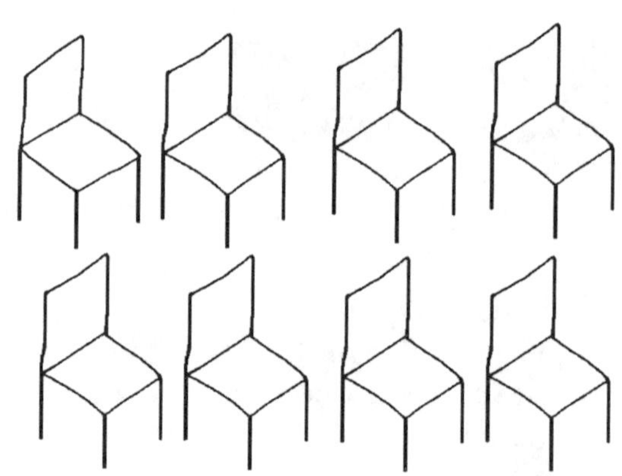

Micha lined up two rows of four chairs. How many rows did Micha make?

Two.

Write: 2 rows of

How many chairs in each row?

Four.

Write: 2 rows of 4

How many chairs did he use?

Eight.

Write: 2 rows of 4 = 8

Briana is choosing people to be on two teams. She needs six people on each team.

How many teams are there?

Two.

Write: 2 teams of

How many on each team?

Six.

Write: 2 teams of 6

How many players on the teams altogether?

Twelve.

Write: 2 teams of 6 = 12

Provide many opportunities for the children to record in this way using words before you introduce the multiplication sign.

2–7 Introducing the Multiplication Sign

Whole-Class or Small-Group Activity

Materials: Classroom objects or connecting cubes, wooden cubes, Color Tiles, or collections

Present multiplication story problems for the children to act out as before, using either real objects or counters. As a story is being acted out, write the equation using words as described in activity 2–6. For example:

Maria's father was making pancakes for three people. He stacked four pancakes on each plate. How many pancakes did he make?

Write: 3 stacks of 4 = 12

Tell the children that there is another way to write this word problem. Erase the words *stacks of* and substitute the multiplication sign for them. Say "stacks of" as you write the sign.

Write: $3 \times 4 = 12$

NOTE: Some children will recognize the sign and call it "times." Acknowledge that it is the "times" sign and that it means "stacks of" in this sentence. Do not allow the children to read the sign as *times,* but also do not say that times is incorrect; just tell them that they need to learn to read it using the words it stands for.

Repeat the activity, presenting more story problems using different kinds of multiplication phrases *(groups of, rows of, piles of, stacks of)*. Each time after you write the equation in words, replace the phrase with the multiplication sign. Have the children read the signs using the appropriate phrases.

2–8 Interpreting Symbols

Whole-Class or Small-Group Activity

Materials: Counters

Write multiplication problems on the chalkboard using numerals and the multiplication sign. Ask the children to act out the problems using counters.

At first, have one of the children choose the words they will use for the multiplication sign (for example, *rows of, stacks of,* or *groups of*). Then have children build the appropriate model. For example, write: 5×3.

Let's make stacks this time. How many stacks will you make? How many in each stack?

Variation: Let each child choose her or his own way of interpreting the problem and share it with the others. For example, write: $4 \times 3 = 12$.

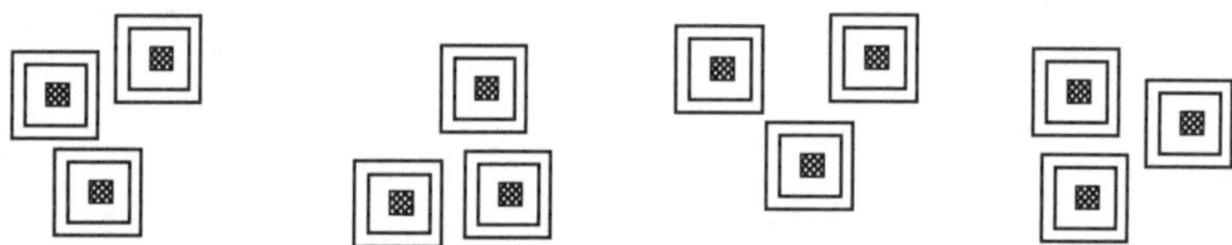

Four groups of three is twelve.

Four rows of three is twelve.

My brother had a birthday. He is twelve years old. His cake had four rows of three candles.

2–9 Acting Out Stories To Go with Multiplication Problems

Whole-Class or Small-Group Activity

Materials: Classroom objects or connecting cubes, wooden cubes, or Color Tiles

Use numerals and the multiplication sign to write a problem on the board and have the children describe a situation that goes with it. For example, write: $4 \times 3 = 12$.

What story can you tell for this problem?

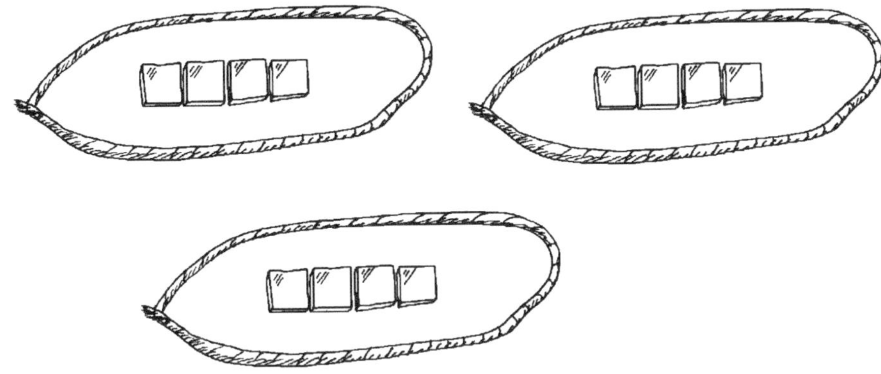

A farmer had three pens. There were four pigs in each pen.

There are three people in my family. Each person got four cookies.

For a related independent activity, see Writing Stories To Go with Multiplication Problems (activity 2–22).

2–10 Learning To Write the Multiplication Sign

Whole-Class or Small-Group Activity

Materials: Small chalkboard, chalk, and eraser or paper and pencil for each child

After the children have had several experiences seeing you write multiplication equations and so are able to read and interpret the equations in a variety of ways, they can begin learning to write the equations themselves.

Level 1: Copying Equations

The children copy the equations that describe various multiplication activities as you write them on the board.

Level 2: Writing Equations and Checking

The children write equations to describe various multiplication situations without any written clues from you. After the children finish writing each of their equations, write it on the board so that they can check their work.

Independent Activities

2–11 How Many Cups? ....Independent Activity

Materials: Counters ▪ Paper or plastic cups (at least ten of the 16-oz size) ▪ 1–6 Number cubes (1 per child) ▪ How Many Groups? Worksheets [BLM #135]

The child rolls a number cube to determine the number of cups to use. After taking that number of cups, the child fills each cup with any number of counters, as long as each cup holds the same number. (In other words, the number of groups is determined by the roll and the number of counters in each group is determined by the child.) The child determines how many counters there are altogether and then records the information on the How Many Groups? Worksheet.

I rolled two. I have to get two cups. I want to put five cubes in each cup.

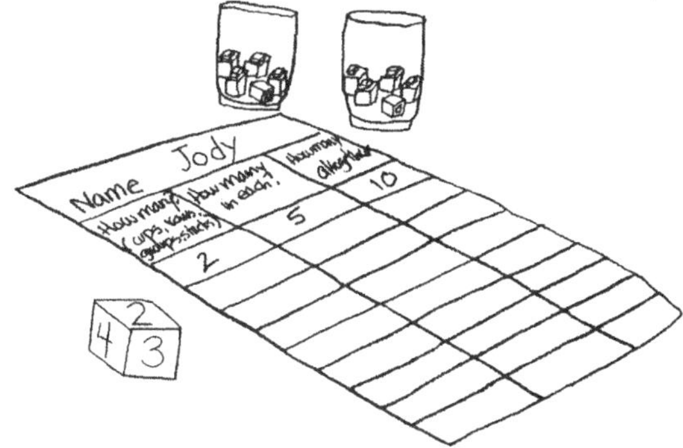

Now I have ten cubes altogether.

I rolled a four. This time I'll get four cups. I'm going to put one cube in each cup.

Now I have four cubes altogether.

2–12 How Many Groups?

Independent Activity

Materials: Counters ▪ 1–6 Number cubes (1 per child) ▪ How Many Groups? Worksheets [BLM #135]

The child rolls a number cube to determine how many groups to make. Then the child decides how many counters to have in each group. A group may have any number of counters in it, but all the groups must be equal. The child records the information on the How Many Groups? Worksheet.

I rolled a three. I need to make three groups. I am going to put four in each group.

Now I have twelve altogether.

2–13 How Many Rows?

Independent Activity

Materials: Counters ▪ 1–6 Number cubes (1 per child) ▪ How Many Groups? Worksheets [BLM #135]

The child rolls a number cube to determine the number of rows to make. Using the counters, the child makes that number of rows of equal length (having first decided on how many to have in each row) and records the information on the worksheet.

I rolled a five.
So I must make five rows.
I will put three in each row.

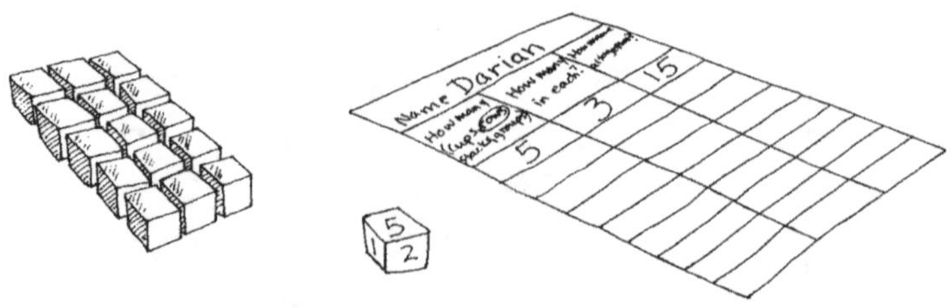

That's 15 altogether.

2–14 How Many Towers? .... Independent Activity

Materials: Connecting cubes, wooden cubes ▪ How Many Groups? Worksheets [BLM #135]

The child rolls a number cube to determine how many towers of cubes to make. A tower may be made from any number of cubes, but all the towers must be equal. The child records the information on the worksheet.

I rolled a two. I can make two towers. I think I'll make them four cubes high.

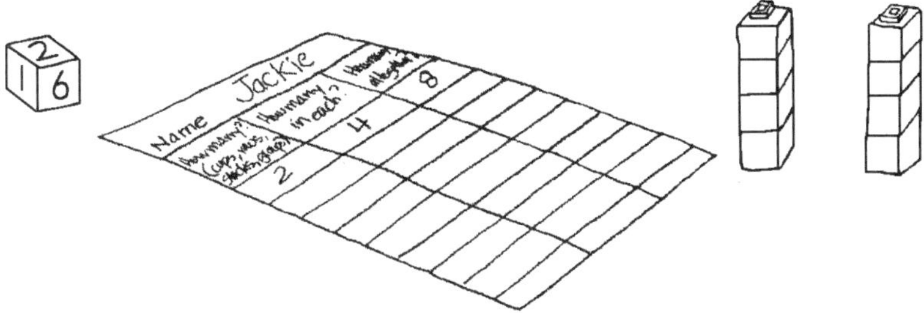

Now I have eight altogether.

2–15 Counting Boards: Multiplication .... Independent Activity

Materials: *Level 1:* Connecting cubes (sorted by color) ▪ Counting boards (1 set of eight identical boards for each child) [BLMs #2–6] ▪ Multiplication Cards [BLMs #137–138]
Level 2: Same as for Level 1, plus 2" × 6" strips of paper (8 per child)
Level 3: Same as for Level 2, but without the Multiplication Cards

The children use the counting boards along with the multiplication cards to practice working with multiplication at a variety of levels.

Level 1: Interpreting Multiplication Problems

Each child chooses a set of eight counting boards, a container of cubes, and eight multiplication cards. The child spreads out the boards and puts a card below each. After deciding what the cubes will represent, the child places them on the boards in rows to represent the problems shown on the cards.

Variation: The child uses several counting boards to display the problem shown on a single card.

Level 2: **Copying and Solving Problems**

The child proceeds as for Level 1 and then writes each problem and its answer on a 2" × 6" strip. After the child completes a series of problems, the papers can be stapled together to form a small book.

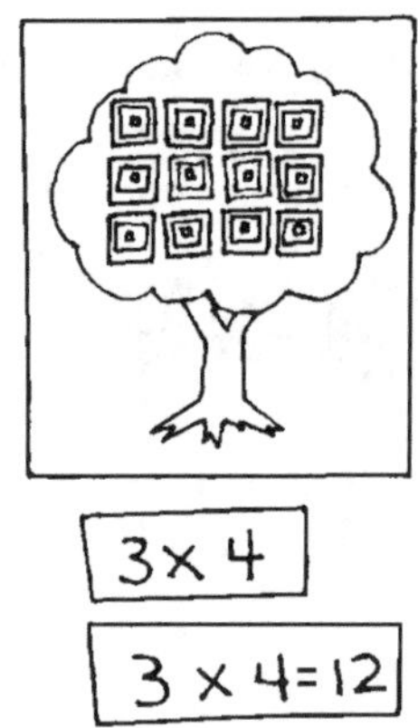

Variation: The child uses as many counting boards as necessary to display a single problem and then writes the answer on a 2" × 6" strip.

Level 3: **Making Up and Solving Problems**

Now, instead of using multiplication cards to indicate the problems, the child makes up his or her own problems, represents them on the counting boards, and writes them on 2" × 6" strips.

Variation: The child uses more than one counting board to display a single problem that he or she makes up and writes on a 2" × 6" strip.

2–16 Problems for Partners: Multiplication .... Independent Activity for Partners

Materials: Counters • Blank paper (1 sheet per child) • Small chalkboard, chalk, and eraser or paper and pencil for each child

One player makes up a multiplication problem and models it by putting groups of counters on a sheet of paper. The other player writes the appropriate multiplication problem and answer in equation form. Then the two players exchange roles. For example:

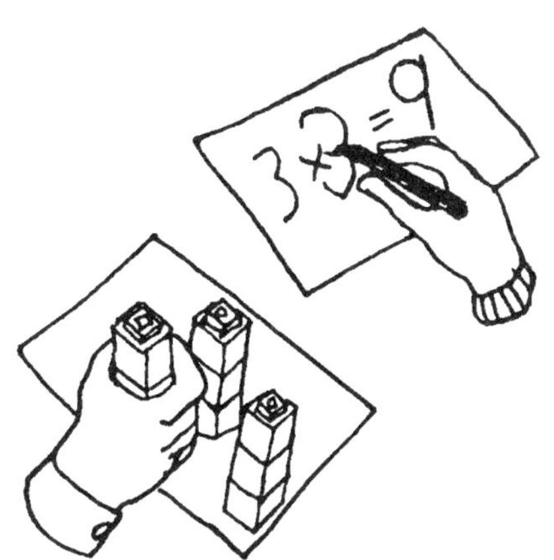

2-17 Roll and Multiply

Independent Activity

Materials: Connecting cubes, wooden cubes, Color Tiles, or collections • Number cubes (1 per child) • Multiplying Strips Worksheets [BLM #136]
Preparation: To make multiplying strips, duplicate BLM #136 and cut it into three vertical strips of five problems each. (Note that the strips can be used repeatedly because new problems will be created each time the children roll the number cubes.)

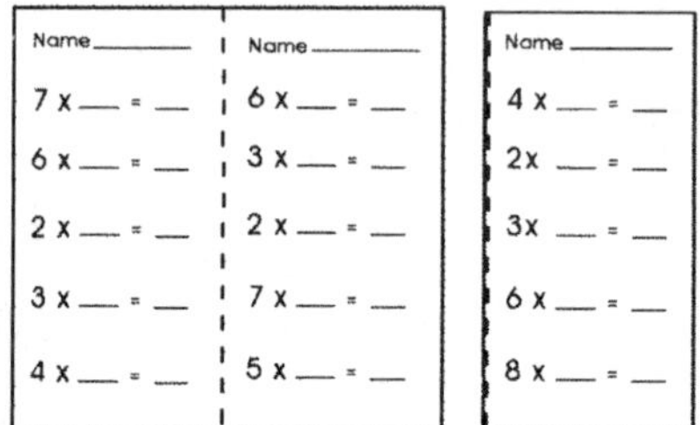

The child rolls the number cube to determine the second number for a problem on the multiplication strip. Then he or she models the problem with counters and records the answer on the strip. Children may complete as many strips as time allows. For example:

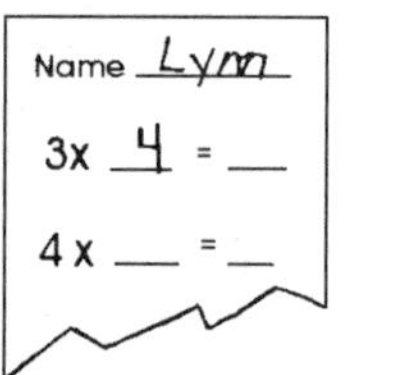

I rolled a four. I wrote a four on my strip.

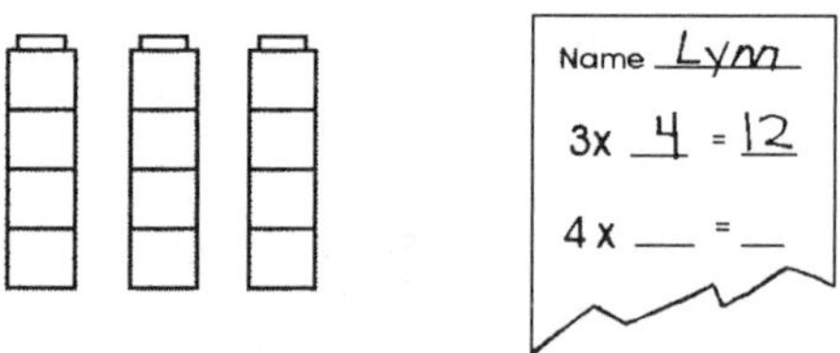

I built three groups of four. That makes twelve. I wrote twelve on my strip.

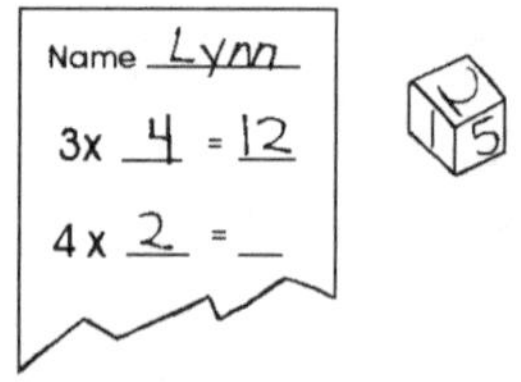

This time I rolled a two.

The child continues rolling the number cube and completing multiplication sentences until one or more strips are completed.

2–18 Discovering Patterns: Cupfuls .... Independent Activity

Materials: *Level 1:* Counters ▪ Plastic cups (at least 10) ▪ 1–6 Number cubes (optional) ▪ Patterns in Multiplying Worksheets (Constant Groups) [BLM #139]
Level 2: Same as for Level 1, plus Place-Value Strips [BLM #113]
Variation: Same as for Level 1, plus 00–99 Charts (1 per child) [BLM #116]

Level 1: Multiplying

The child picks any number of cups or rolls a number cube to determine the number of cups to use for the activity. The child writes the number in the first blank in each of the ten problems on the Patterns in Multiplying Worksheet, uses the counters and cups to model the problems, and then writes the answers.

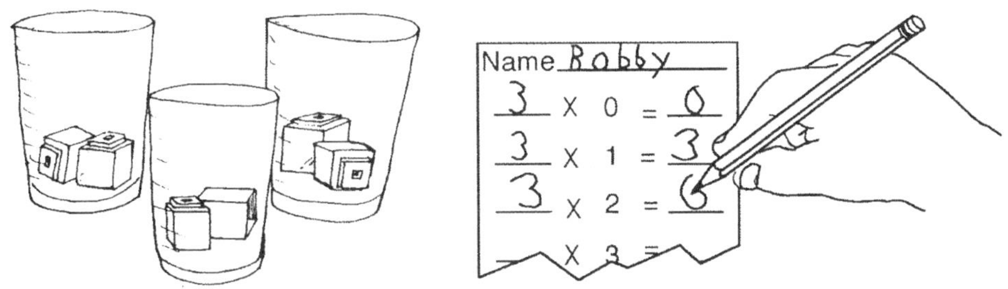

I worked with four cups yesterday. Today I am using three cups.

Level 2: Looking for Patterns

Because the multiplication equations appear on the worksheet in numerical order, children can use their work to look for patterns.

After the children finish the ten equations, they transfer their answers, in order, onto a place-value strip. They loop the patterns they find in each column and then, without using the counters, they extend the pattern on the strip.

Variation: The children color their answers on the 00–99 chart and identify the pattern that emerges. Then they extend the pattern without using counters.

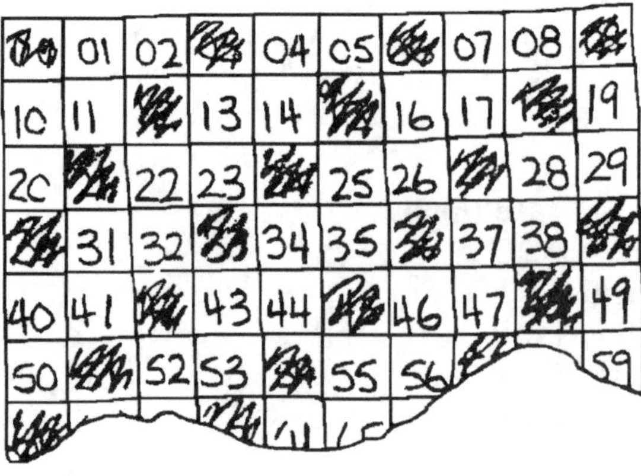

2–19 Discovering Patterns: Buildings

Independent Activity

Materials: Connecting cubes ▪ 1–6 Number cubes (optional) ▪ Patterns in Multiplying Worksheets (Sequential Groups) [BLM #140] ▪ Place-Value Strips [BLM #113]
Variation: Same as for Levels 1 and 2, plus 00–99 Charts [BLM #116]

Level 1: **Multiplying**

The child picks a number (or rolls a number cube to determine a number) and then makes a stack or "building" with that number of cubes. The child keeps on making buildings, adding one at a time to model each problem on the Patterns in Multiplying Worksheet and then writes the answer for each.

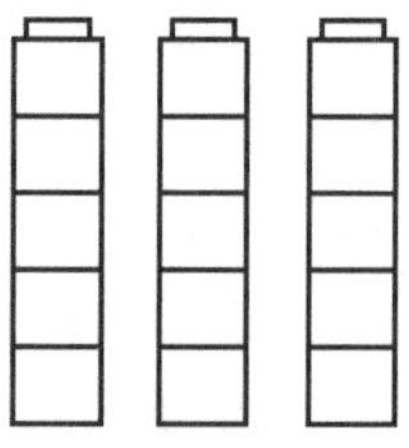

Name Nan
0 X 5 = 0
1 X 5 = 5
2 X 5 = 10
3 X 5 = 15

Level 2: Looking for Patterns

The child transfers each of the answers, in order, from the finished worksheet onto a place-value strip. Then the child loops the patterns that he or she sees on the strip and extends them without using the cubes.

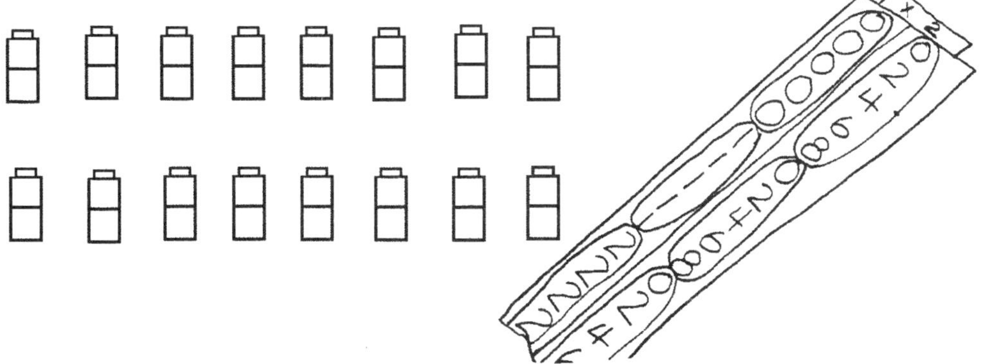

Variation: The children color in the answers on a 00–99 chart, look for the pattern that emerges, and then extend the pattern without using the cubes.

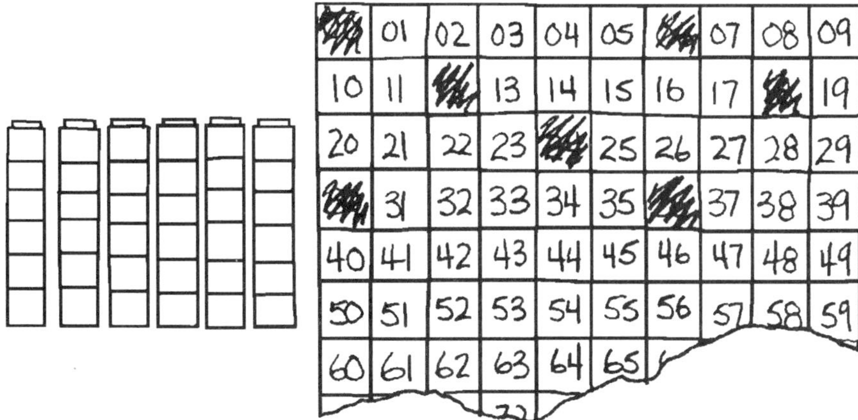

2–20 Discovering Patterns: Number Shapes

Independent Activity

Materials: Counters ▪ Number cubes (optional) ▪ Number Shapes (Make at least nine of each available.) [BLMs #76–82]
Level 1: Patterns in Multiplying Worksheets (Sequential Groups) [BLM #140]
Level 2: Place-Value Strips [BLM #113]
Variation: Same as for Levels 1 and 2, plus 00–99 Charts [BLM #116]

Level 1: Multiplying

The child chooses one set of number shapes and writes the number of squares that make up that shape in the first blank on the worksheet. Then the child uses the counters on the shapes to find the products and records them on the worksheet.

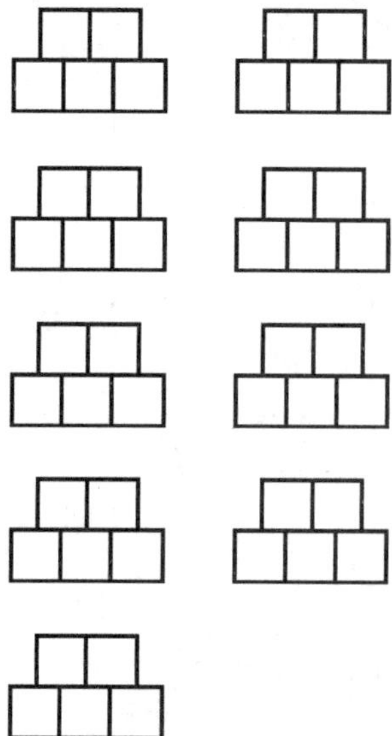

I used the four number shape yesterday. Today I am using the five shape.

Name Emily
0 X 5 = 0
1 X 5 = 5
2 X 5 = 10
3 X 5 = 15
4 X 5 = 20
5 X 5 = 25
6 X 5 = 30
7 X 5 = 35
8 X 5 = 40
9 X 5 = 45

Level 2: Looking for Patterns

The child transfers each of the answers on the completed worksheet, in order, onto a place-value strip. Then the child loops the patterns and extends it without using counters and number shapes.

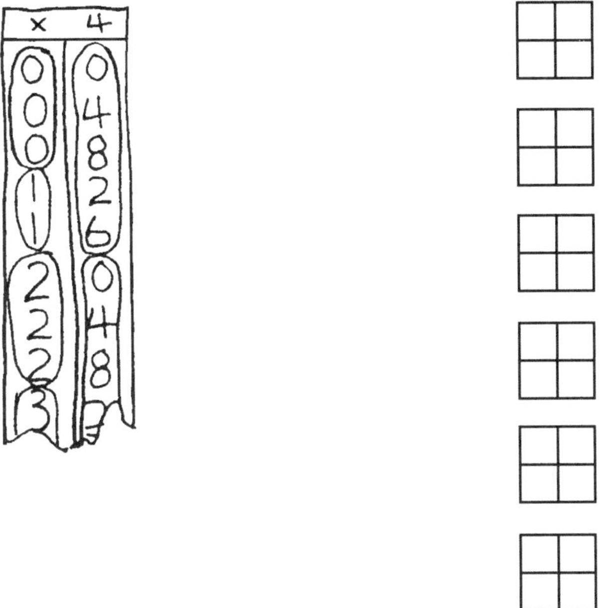

Variation: The children color their answers on a 00–99 chart, look for the pattern that emerges, and then extend the pattern without using counters.

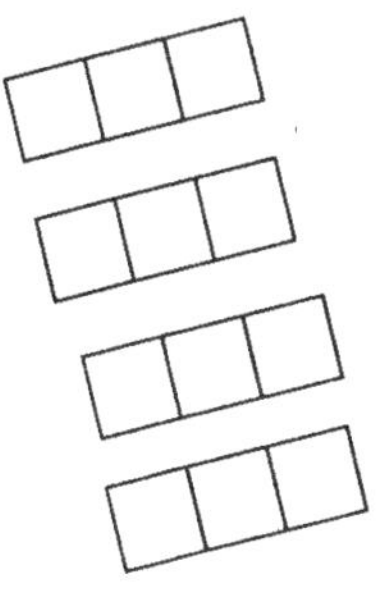

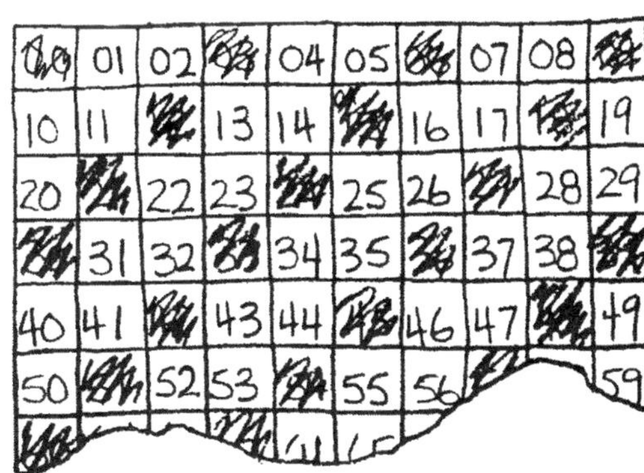

2–21 Pattern Search: Multiplication

Independent Activity

Materials: 12" × 18" construction paper • 00–99 Charts (1 per child for each pattern they explore) [BLM #116] • Crayons

This is an adaptation of Introducing Pattern Searches (activity 1–16). It may be a new activity for some children and an extension for others. The essential difference is that, in this context, when children identify patterns they record them in terms of multiplication. The children explore, identify, and then sort patterns according to the class's agreed upon colors, as described in Naming Patterns with Colors (activity 1–13).

Throughout this ongoing activity, children discover patterns that occur and reoccur in various situations. They will begin to see connections between patterns and notice when familiar patterns emerge in new and different situations. Experiencing the same patterns over and over again and becoming familiar with them will help the children recognize these patterns as they work with multiplication.

For this activity, the children search for patterns formed by groups of things in their environment. They record the patterns on a T-table and as multiplication equations. For example:

One hand has how many fingers?

Five.

Hands	Fingers	
1	5	1 × 5 = 5

Two hands have how many fingers?

Ten.

Hands	Fingers	
1	5	1 × 5 = 5
2	10	2 × 5 = 10

Three hands have how many?

Fifteen.

Hands	Fingers	
1	5	1 × 5 = 5
2	10	2 × 5 = 10
3	15	3 × 5 = 15

What pattern do we find when we count the fingers on our hands?

Supply 12" × 18" paper on which the children can record their patterns. As was suggested for activity 1–16, children should include a picture, a T-table, and a 00–99 chart marked to show the pattern.

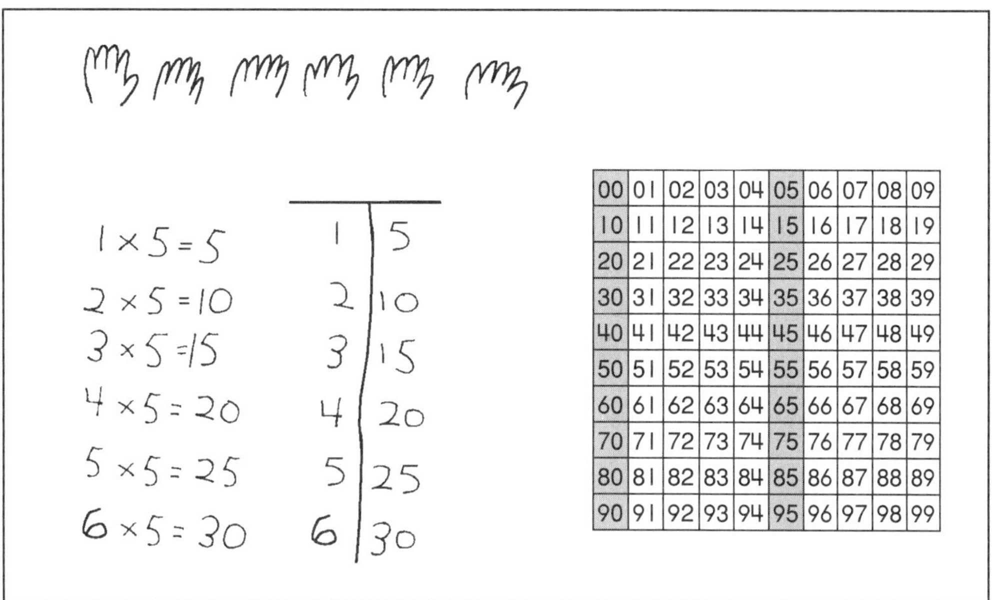

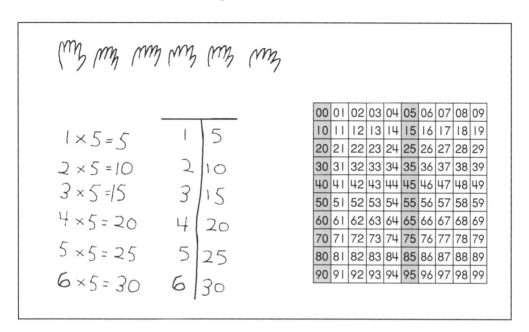

Have the children explore a variety of patterns and record the multiplication equations that describe them.

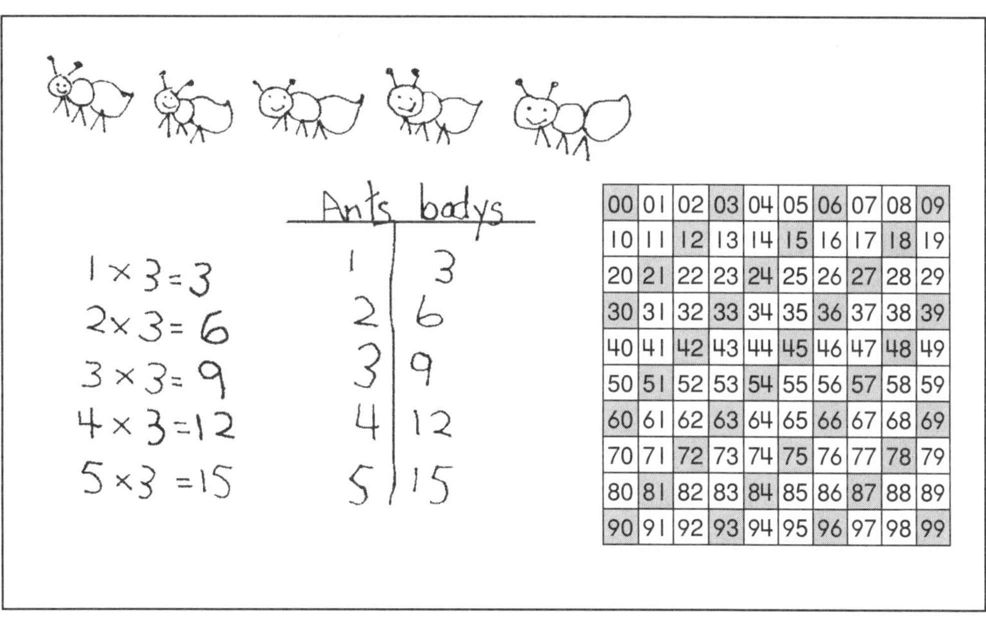

2–22 Writing Stories To Go with Multiplication Problems .... Independent Activity

Materials: Multiplication Cards [BLMs #137–138] • Writing paper and pencil and crayons for each child

The children write a story problem for each multiplication equation. There are several ways the equation can be determined:

- Write an equation on the board so that the whole class works from the same one.
- Children may draw from a pool of multiplication cards.
- Children may think of an equation for themselves.

For example:

$3 \times 4 = 12$

The old woman who lived in the shoe had three beds for her children. Four children slept in each bed.

$3 \times 2 = 6$

I have three horses. I fed each of my horses two carrots.

$6 \times 5 = 30$

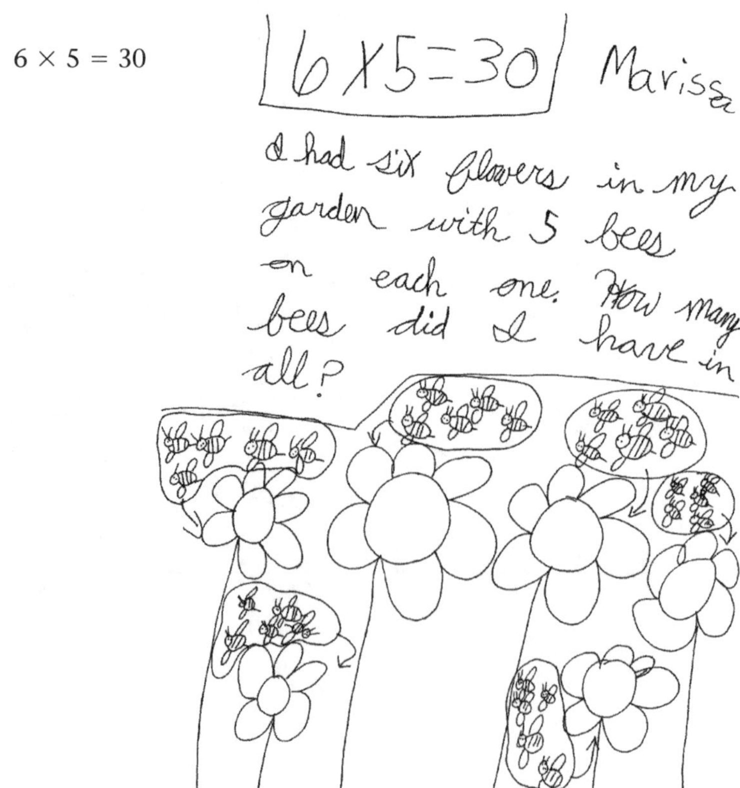

6X5=30
Marissa
I had six flowers in my garden with 5 bees on each one. How many bees did I have in all?

2–23 Lots of Rectangles .... Independent Activity

Materials: Connecting cubes, wooden cubes, or Color Tiles • Rectangle Cards (See examples below.) • 2" × 6" strips of paper
Preparation: Draw a variety of rectangles on cards that measure approximately 6" × 9".

The child selects a card, fills the rectangular shape with cubes or tiles, and writes the multiplication equation that describes the array that is formed. For example:

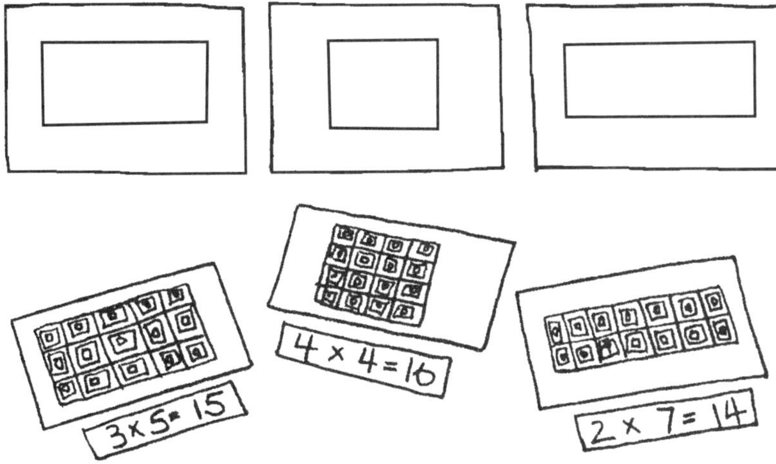

2–24 Shape Puzzles: Multiplication .... Independent Activity

Materials: Connecting cubes, wooden cubes, or Color Tiles • Shape Puzzles [BLMs #35–39] • 2" × 6" strips of paper
Extension: See Paper Shapes (activity 1–33) for directions for making shapes from tagboard.

The shape puzzles used for counting and for addition and subtraction practice can also be used for multiplication, providing a challenge for those children who are ready. With those puzzles, the children can learn to describe arrangements of cubes as *combinations of multiplication and addition.* In this context, you can introduce the use of parentheses in equations. For example:

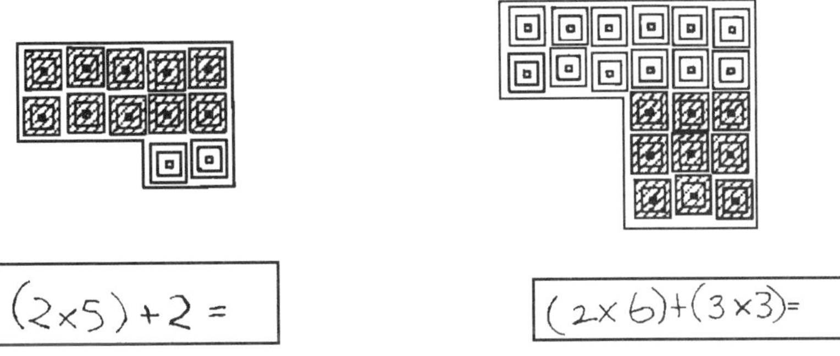

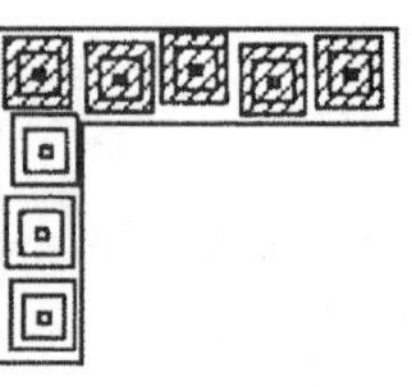

(1×5) + 3 =

Extension: Larger Numbers

If the children are ready to work with larger numbers, they can use the larger paper shapes used for studying place value. (See activity 1–33 for examples of paper shapes.)

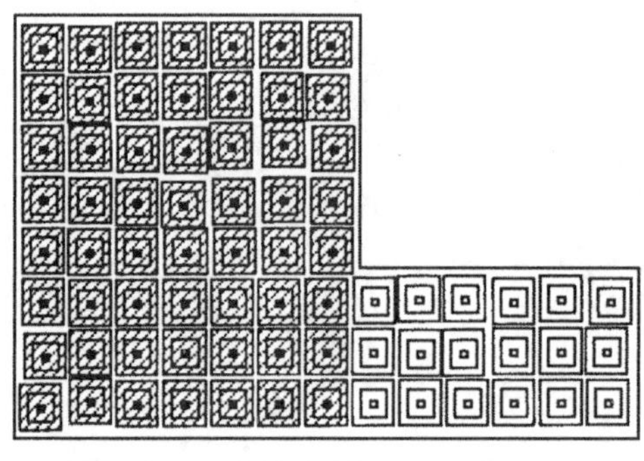

(7×8) + (3×6) =

(3×5) + (4×2) + (4×2) =

or

(3×5) + (4×4) =

2–25 Mixing Them Up .... Independent Activity for Partners

Materials: Counters ▪ Addition Cards [BLMs #64–68], Subtraction Cards [BLMs #69–73], Multiplication Cards [BLMs #137–138] ▪ Use a More-or-Less Spinner if you have one. (See Book 1, p. 145.) Otherwise, make a More-or-Less Die by marking three faces of a wooden cube with "More" and three faces with "Less."

The following game can be a challenge for two players who are competent in adding, subtracting, and multiplying.

Mix up a pool of addition, subtraction, and multiplication problem cards. To start the game, each player draws one card from the pool.

4 X 2		10 - 4

The players use counters to model the problems indicated on their cards and determine the answers.

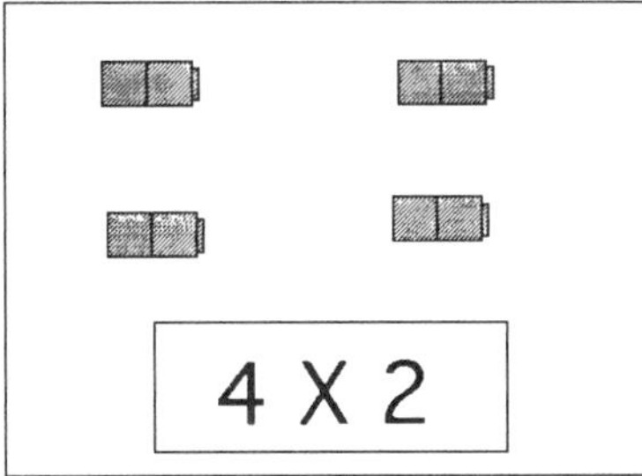

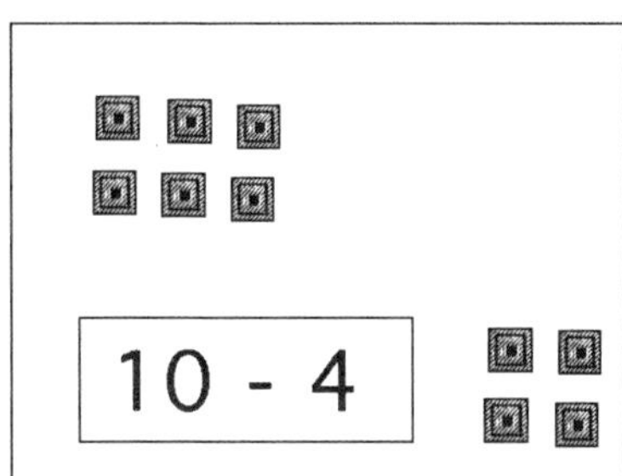

One player then spins the More-or-Less Spinner to determine who wins the two cards—the player with more or the player with less. For example:

***Less* wins, so I keep both cards.**

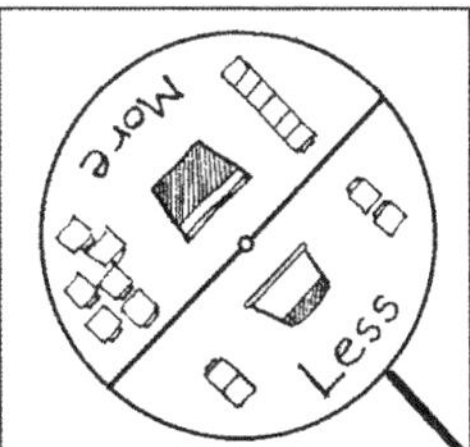

Players continue picking cards and spinning the spinner until all the problem cards have been won or until time is up.

When textbook or curriculum objectives are:

- Understanding the process of division
  - Dividing with single-digit divisors

8 ÷ 2

Then you want to teach

Beginning Division

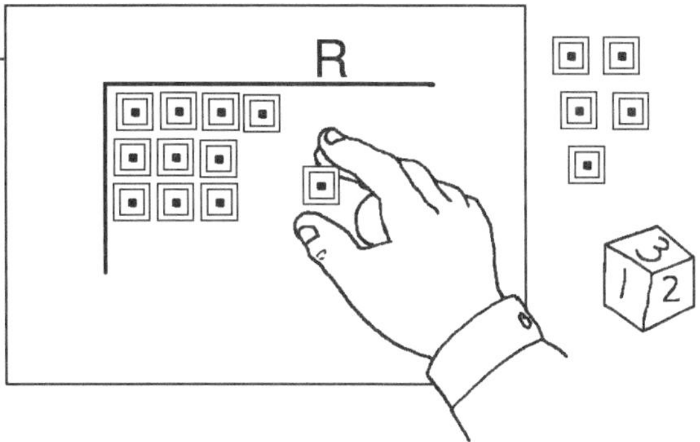

What You Need to Know About Beginning Division

Division is a natural process for children because it relates to their experiences of sharing.

Division is a process that children deal with naturally in their everyday lives in situations such as sharing cookies or forming teams. It is important to help children relate this familiar world to the world of written problems. The activities in this chapter give children formal experiences with division. They learn to interpret the language of division and to explore the number patterns and relationships that occur when numbers are divided.

Two types of situations call for the operation of division: the *grouping* process (determining how many groups) and the *sharing* process (determining how many in each group).

There are two types of division: sharing and grouping.

In the grouping process, we divide a given quantity of objects into smaller groups of a particular size to determine the number of groups that can be made. For example:

> We have fifteen children in our singing group. Five of them can sit next to each other on the rug. How many rows of children can sit on the rug?
>
> Snap together seventeen cubes. Now break them up into buildings that are each four cubes tall. How many buildings can you make? Are there any cubes left over?

In the sharing process, we divide a quantity of objects into a particular number of groups to determine the number of objects in each group. For example:

> We have fourteen chairs. We need to put them into two rows of the same length. How many chairs do we have in each row? Do we have any leftovers?
>
> Get nineteen cubes. Divide those cubes into three equal piles. How many cubes are in each pile? Are there any left over?

The differences between the two processes are subtle, and it is not necessary to teach children to distinguish between them or learn to label them. The point is to provide many experiences with both types of process so the children are comfortable with each one.

To the adult it may seem that children should easily see the relationship between multiplication and division. However, to young children who are just beginning to work with these concepts, the two operations appear to be entirely different. It will challenge some children simply to interpret the situations appropriately without also considering how they relate to each other. Provide opportunities for children to work with problems that are related, but don't rush them to see the relationship. The more at ease they are interpreting the language of multiplication and division and in solving problems, the more likely it is that they will discover the relationship for themselves.

The relationship between multiplication and division is not obvious to young children.

Sometimes we teach beginning division using only problems with no remainders—problems with answers that "come out even." Remainders are a natural and common part of the division process, so it serves children to recognize this and to learn to deal with remainders right away.

Children should deal with the concept of remainders from the beginning of their work with division.

Children develop mathematical ideas over time. Provide them with many opportunities to work with division to develop these ideas in a variety of ways.

Teaching and Learning Beginning Division

Through the activities in this chapter, children develop an understanding of beginning division in ways that help them recognize the process and learn the language of division. The teacher-directed activities give children opportunities to interpret this language, act out division situations, and record their division experiences. The independent activities provide children with the ongoing practice they need in order to become comfortable interpreting division situations and solving division problems.

Using the Chapter

The "Meeting the Needs of Your Children" charts in the introduction to this book and the *Planning Guide* that accompanies this series offer detailed information that can help you plan how to use the chapter's activities. The following are general suggestions for using the activities with different groups of children.

Kindergarten and First Grade The activities in this chapter are not appropriate for kindergarten or first-grade children.

Second Grade Whether or not children are expected to begin working with division in second grade varies from one school district to another. The activities in this chapter are appropriate for use with second-grade children toward the end of the school year.

Third Grade Division is an important concept to be learned by third-grade students. Helping children build a base for understanding division is vital, and so is the focus of the activities in this chapter.

Children with Special Needs If you have special-needs children who need help with division concepts, this chapter can be of help as it deals with very basic ideas of division. See the "Meeting the Needs of Your Children" chart for Chapter 3 in the introduction to this book and take the children as far as it seems appropriate for them.

Goals for Children's Learning*

Goals

When presented with a variety of division situations, children will be able to:

- Recognize both the processes of sharing and grouping (partitioning) as division
- Interpret the language of division as it occurs in story problems

* Adapted from *How Do We Know They're Learning? Assessing Math Concepts.*

- Interpret division situations
- Write division problems to describe appropriate situations
- Solve division problems
- See the relationship between multiplication and division problems

Analyzing and Assessing Children's Needs

When our goal is for children to develop an understanding of the process of division as it occurs in the real world, it is not enough to know if children have memorized division facts or if they can complete worksheet pages. Instead, we need to know if they can recognize the need to divide to solve problems in a variety of settings. The nature of the activities you will use to introduce division enables you to get the information you need by observing the children at work. The following questions can help you recognize the stages through which children move as they work with beginning division.

Questions to Guide Your Observations*

Questions

Interpreting the Language of Division

- Can the children interpret the language of division by modeling how to divide quantities into *rows, groups, stacks,* and *piles* of objects?
- Can they determine when the answer refers to the *number of groups,* versus the *number in each group?*

Interpreting Simple Story Problems

- Can the children interpret division story problems using physical models or drawings?
- Do they interpret the problem with ease or with difficulty? Do they need any prompts or hints?
- Do they choose the method appropriate to the situation?

Interpreting Symbols

- Can the children interpret division equations using models and/or story problems?
- Can they read and interpret equations that use both types of division symbols (÷ and $\overline{)\quad}$)?
- Can they make up a story problem to go with an equation?

* Adapted from *How Do We Know They're Learning? Assessing Math Concepts.*

Writing and Reading Equations

- Can the children write equations to describe story problems? Is this easy for them or is it challenging?
- Can they read the equation back?
- Do they know how the numbers connect to the situation in the story?
- If children do not know the formal way of recording a division story, can they represent the story symbolically in some way?

Solving Division Problems

- When the children solve division problems, what kind of strategies do they use?

  Do they divide quantities by sharing one by one?

  Do they begin sharing by first giving more than one to each group and then giving ones to each group?

  Are they able to chose an appropriate number by which to divide a group?

  Do they know how to solve without having to use models?
- Do they see any relationship between multiplication and division? Can they use their knowledge of multiplication to aid them when dividing?

Meeting the Range of Needs

These beginning division experiences will naturally meet a range of needs because children will approach them in a variety of ways. Some children will work out problems slowly and deliberately. Others will be able to look for relationships and solve problems with ease. The size of the numbers children are asked to work with can also be varied to more accurately meet an individual student's needs.

A Classroom Scene

Mrs. Tanaka introduced beginning division to her class several days ago. She began by having the children act out division story problems. For the first couple of days, her focus was on introducing the division process to the class as a whole. She worked for several days with the teacher-directed activities and then began introducing the independent activities.

When the children work independently at stations, Mrs. Tanaka is able to see how individuals are thinking about the division process. She wants to see how well her children are understanding, but she knows from past experience that she can't watch everyone at once, so she has selected a few students to focus on today.

Independent Activity: *Division Stations*

Over the past several days, Mrs. Tanaka has presented five division stations to the children. She plans to introduce two more tomorrow. She is not going to introduce any new stations today because she wants to give the children plenty of time to work so that she can see how they are doing.

Dino is one of several children who are working with the counting boards and Counting Boards: Division (activity 3–9). Mrs. Tanaka stops to watch Dino for a few minutes.

Dino has chosen a division card and is putting out the counting boards and cubes to represent the problem shown on the card. He seems engrossed and is not having any difficulties at this point. He is very carefully "sharing," or counting out one cube at a time for each board.

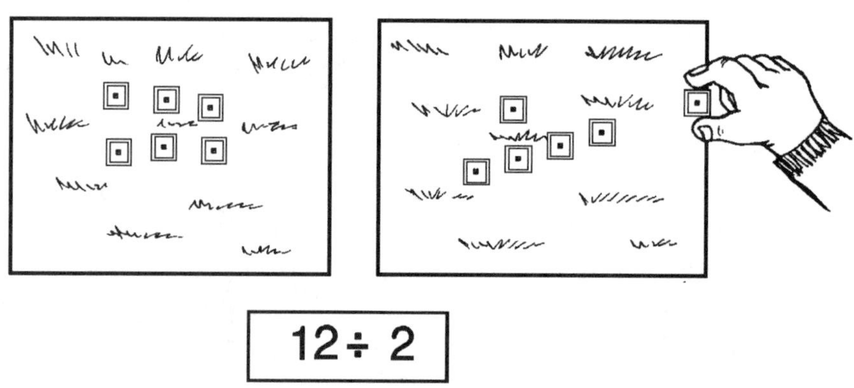

Mrs. Tanaka sees that Matt has completed several problems already.

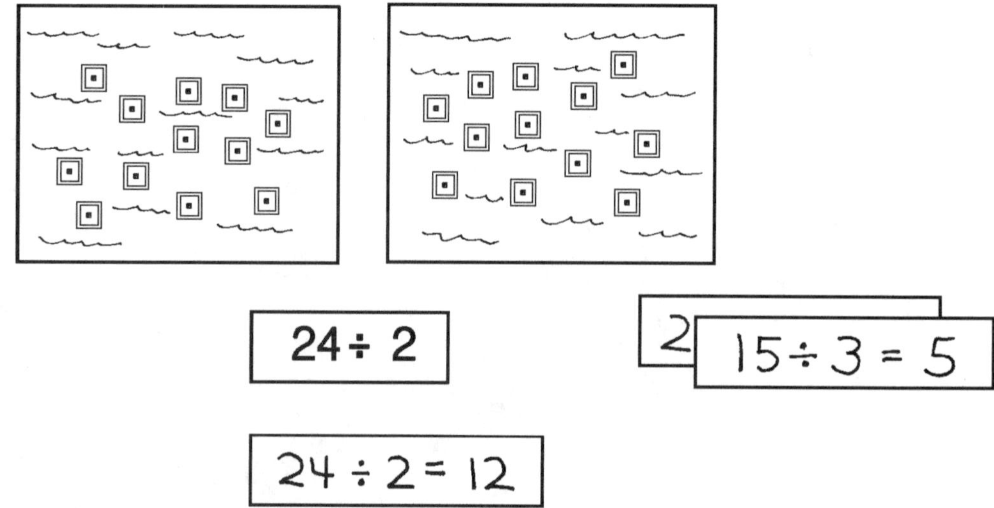

He looks up at Mrs. Tanaka and comments, "This is fun. I try to guess what the answer will be before I really use the cubes. See, I knew 24 divided by two was probably 12."

"How did you figure that out?" Mrs. Tanaka asks.

I don't know. I just knew," Matt replies. He has a strong sense of number but is still working on explaining his thinking.

Mrs. Tanaka doesn't want to slow down Matt's thinking, but she also wants to see if she can help him become more aware of where his answers come from. "I wonder if you knew something else about twelve that helped you in this case. Can you think of something that might have helped you?"

"Well . . . ," says Matt, "I know that 12 and 12 makes 24, so it just seemed like I would have two twelves when I divided, too." Matt understands a lot about numbers, but is as intrigued as are the other children with these division activities because he is still trying to figure out number relationships.

Mrs. Tanaka goes on to observe the children at work with Making Rows (activity 3–11). She wants to see how Joni is doing today. She worked with her for a while yesterday and wants to make sure that she doesn't need additional help.

Joni is lining up rows on the Making Rows card. She started with 16, then rolled a three on a number cube to determine that she had to make three rows. She is placing one cube at a time in each row. She chose this activity yesterday, too. She wasn't sure how to record the process on the worksheet at first, but today she is quite confident.

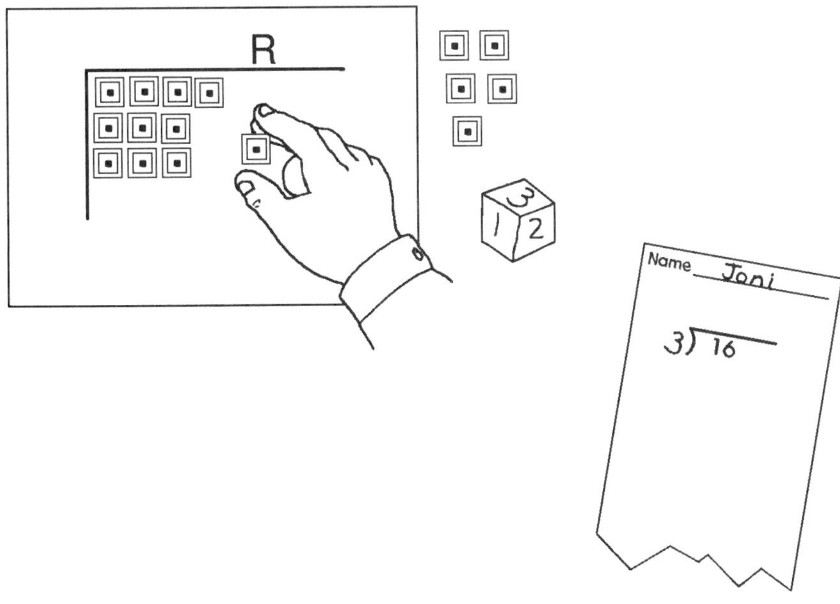

Mrs. Tanaka glances over at Leon, who is working with Cups of Cubes (activity 3–13). Leon has divided 12 acorns among three cups. He recorded the three in the appropriate place on the Dividing Strips Worksheet, but now he seems hesitant to write the answer. "I have four and four and four, so what do I write?"

Mrs. Tanaka explains that he needs to write the four only once because when he writes the number four, it refers to what is in *each* cup.

At the station for Problems for Partners: Division (activity 3–12), Gina and Jake are partners. Gina has presented a problem to Jake. She started by making a train of 12 cubes and then decided to break up the train into lengths of four.

Jake wrote "12 ÷ 4 = 4."

"Look again," Gina tells Jake. "We don't have four trains of four. We have three trains."

Jake looks a little puzzled, so Gina tries to point out the problem again. "See, we had 12, and I broke off as many fours as I could. I could make one, two, three fours," she says, pointing at each of the trains. "So that means twelve divided by four equals three. Now you make a train, and you break it up, and I'll see if I can write it down."

Mrs. Tanaka sees that Jake is still confused, but she is willing to let him try to understand a bit longer before she becomes involved. She knows that certain ideas are difficult for children when they are learning to divide. Many children need to experience the process a few times before overcoming their initial confusion. The teacher will come back in a few minutes to see if Jake is still having difficulty or if division is starting to make sense to him.

As she observes, Mrs. Tanaka realizes that most of the children are working hard to interpret the division problems. She also realizes that they will need to work at these stations for quite a while before they become able to predict some of the answers before having to go through the process of figuring them out. This is natural when children work with a new process. If she allows children the time they need, they will eventually become focused less on how to do the tasks and how to write the appropriate equations and more focused on what is happening with the numbers. So far, Matt is the only child Mrs. Tanaka has observed who is ready and able to think about the relationships among the numbers he is using.

About the Activities

Developing the Concept of Division Through Teacher-Directed Activities

Children must understand that division is a natural process that occurs in the real world. The teacher-directed activities in this chapter help you introduce division through story problems and the building of models. Repeat these activities over and over again so that children become comfortable with the process and develop an increased awareness of what happens to numbers when they are divided.

Practicing Division Through Independent Activities

The independent activities provide opportunities for children to practice division. They can do these activities alone or with partners and without direct teacher involvement. Introduce these activities during teacher-directed lessons to be sure that the children know what to do for each activity and that, when required, they know how to write division equations with confidence.

Teacher-Directed Activities

3–1 Acting Out Division Stories: Using Real Objects

Whole-Class or Small-Group Activity

Materials: Groups of classroom objects

Tell story problems to the children that will help them focus on the process of division. Have the children act out your stories using real classroom objects. Be sure to include some problems that involve remainders. Remainders are a natural part of dividing, and so children should deal with them as they begin to learn how to divide. For example:

Angela has twenty-five pieces of paper to hand out for booklet covers. Each child needs two pieces of paper to make a cover. How many children can have two pieces?

Lorenzo has sixteen pencils. He is going to divide them among five of his friends. How many pencils will each friend get?

There are fourteen books on the shelf. If each child takes three of them, how many children will get three books?

There are eight bottles of glue. Each table of children will get two of those bottles. How many tables will get glue?

NOTE: Let the children deal with the concept of "leftovers," or remainders, beginning with these very first experiences.

3–2 Acting Out Division Stories: Using Counters .... Whole-Class or Small-Group Activity

Materials: Counters ▪ Blank paper or 9" × 12" construction paper (1 sheet per child)

You can present a wide variety of familiar, real-world situations by having children use counters to act out your division story problems. Cubes, Color Tiles, buttons—any type of counter—can be used to represent people, animals, or objects. For example:

Joan collects stamps. She has twenty-two stamps. Four stamps fit on each page of her stamp-collection book. How many pages can she fill?

Twenty-five boys and girls signed up to play basketball. They need five players on each team. How many teams can they make?

Fifteen children are in line to ride the Ferris wheel. Three of them can fit in each seat. How many seats will they use?

Andy has eighteen wheels in his building set. How many cars can he make using all the wheels?

Becky has six friends visiting her. Her mother said that she and her friends could share a package of twenty-four cookies. How many cookies will each child get?

(For the last problem, notice whether or not the children divide by seven, to see if they remembered to include Becky as one of those who will get cookies.)

Extension: Sharing Remainders

Challenge the children to figure out how to share remainders when it is appropriate to do so. For example:

Steven had seven date bars. He shared them with his two little brothers. How many date bars did Steven and his brothers each get?

Also ask children to determine when a remainder can and cannot be shared. For example:

Linda had 7 balloons. She wants to share them with her friend Kathy. Will there be any leftovers? Can the girls share the leftovers? Why or why not?

3–3 Building Models of Division Problems*

Whole-Class or Small-Group Activity

Materials: Connecting cubes, wooden cubes, Color Tiles, or collections ▪ Plastic cups (16 oz)

You can help children become comfortable dividing objects into equal groups at the same time you explore numbers with them. The following are examples of directions that will lead the children to build models for division problems. You will need to repeat these kinds of directions in many lessons so that children can become comfortable with the language of division.

Stacks

Make a train twenty-four cubes long. How many stacks of three can you make from this train?

Put the train back together. Now find out how many stacks of two you can make from it.

Divide your twenty-four cubes into four stacks. Make each stack the same height.

Now divide the twenty-four cubes into three stacks all the same size. How many are in each stack?

Rows

Get sixteen tiles. How many rows of eight can you make using all these tiles?

How many rows can you make if each row is six tiles long?

Make two rows using all sixteen tiles. How many tiles are in each row?

Now divide the sixteen tiles into five rows all the same length. How many tiles are in each row?

Groups

Get three plastic cups. Divide twelve counters among these cups so that you have the same number of counters in each cup. How many counters are in each cup?

Now get sixteen counters. Put them into the three cups so that you have the same number in each cup. How many counters go into each cup?

Get fifteen counters. Put them into two cups so that you have the same number in both cups. How many counters are in each cup? Any leftovers?

Now put fifteen counters into three cups. How many counters are in each cup?

How many cups should we try this time? How many counters should we use?

* Based on *Mathematics Their Way,* "Unifix Trains," p. 202.

3–4 Odds and Evens .... Whole-Class or Small-Group Activity

Materials: Connecting cubes, collections ▪ 00–99 Charts (1 per child) [BLM #116]

Help children discover the pattern of the odd and even numbers by exploring a variety of numbers with them to see which will divide evenly. Record the results on the 00–99 charts. For example:

I have a train of seven cubes. When I try to break up this train into two trains that are the same length, I have one odd cube sticking out.

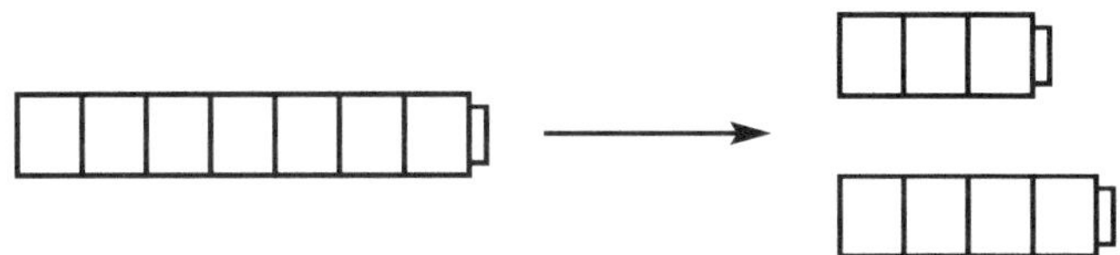

When we try and divide a number of things into two equal parts and one odd thing is left over, we call the number an odd number.

Let's try another number to see if it is odd. Get sixteen cubes and snap them together to form a train. Then try to break up the train into two trains that are the same length.

Did they come out the same?

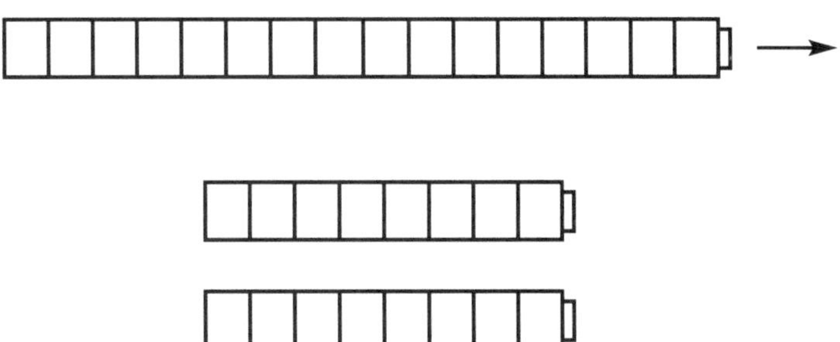

Yes. We have eight and eight.

When we can divide a number into two equal parts and there are no leftovers, we call that number an even number.

Let's use the 00–99 chart to mark the numbers as we discover whether they are even or odd. We'll circle the even numbers and put an X on each odd number.

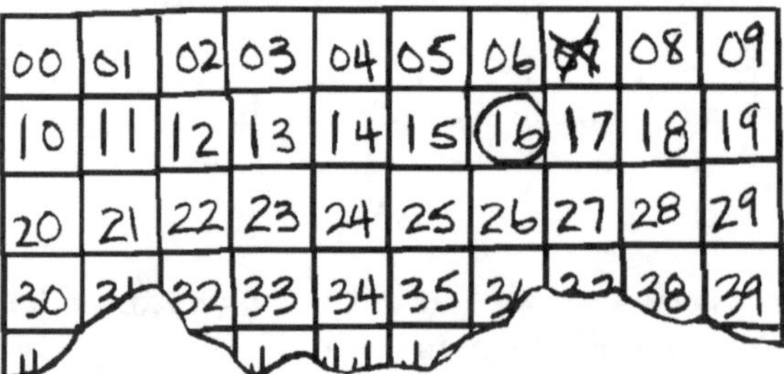

What number should we try now?

Fourteen.

What happens when we try to divide fourteen into two equal trains?

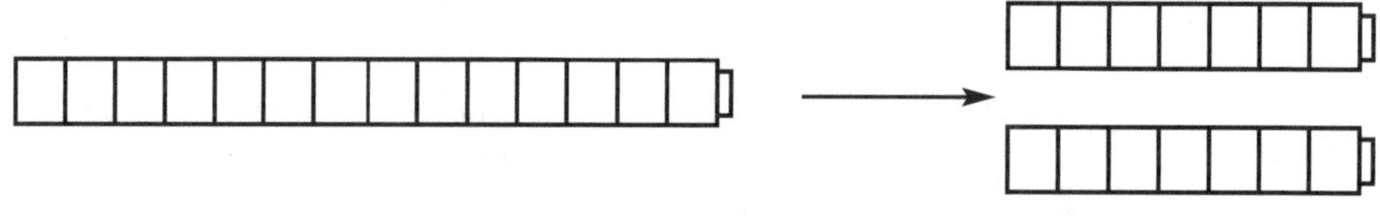

There's no odd one. So, fourteen's an even number.

Continue to help children explore different numbers. Once the children see the pattern emerging on the 00–99 chart, have them first predict whether or not a train made for any particular number can be broken up evenly and then have them check to see if that number is odd or even. On another day, have children build their models from collections instead of cubes. This will provide them with a greater challenge because the "odd one left over" won't be so obvious.

Variation: Help children explore the numbers in a different way to find out if the numbers are odd or even. This time, have the children break up each given number of cubes into short trains of two cubes each. For example:

Let's break up our train of eight cubes into twos.

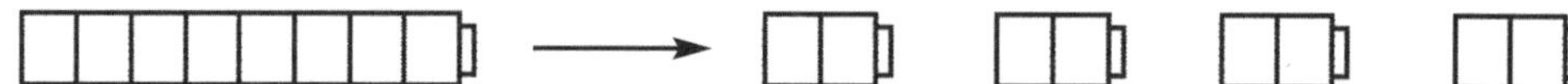

Do we have any odd ones left?

No.

Let's try nine.

Are there any odd ones left?

Yes, we have one odd one.

3–5 Modeling the Recording of Division Experiences

Whole-Class or Small-Group Activity

Materials: Classroom objects or counters

When the children can interpret the language of division with ease, show them how to record the division process using symbols.

Continue to tell division story problems and have the children act them out either with actual classroom objects or counters. As they do this, write the problems symbolically on the chalkboard. For example:

Jerry got eight books.

He gave two books each to as many people as he could.

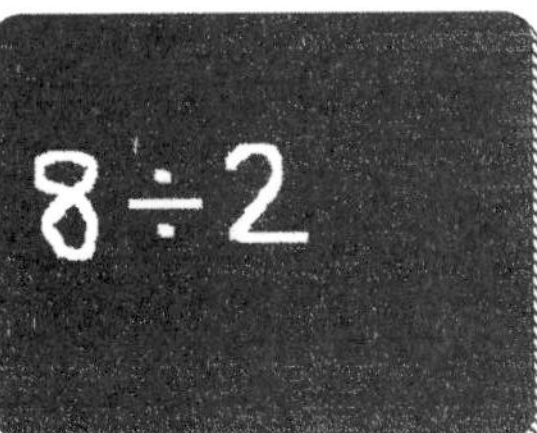

How many people got two books?

Four people got two books.

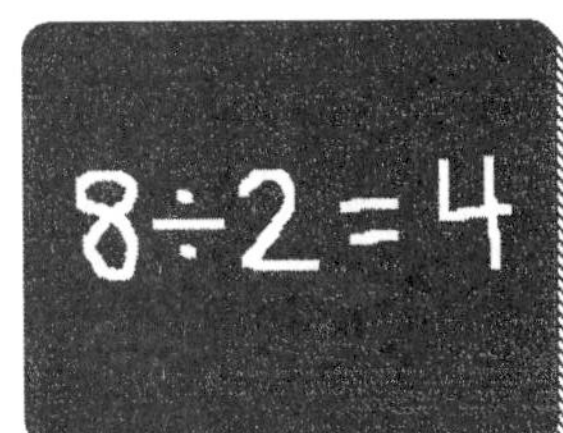

Cecelia got eleven books. She gave the same number of books to each of two people. How many books did each person get?

Eleven ... divided by two ... is five. One is left over.

On other occasions, show the children the other way to write division problems.

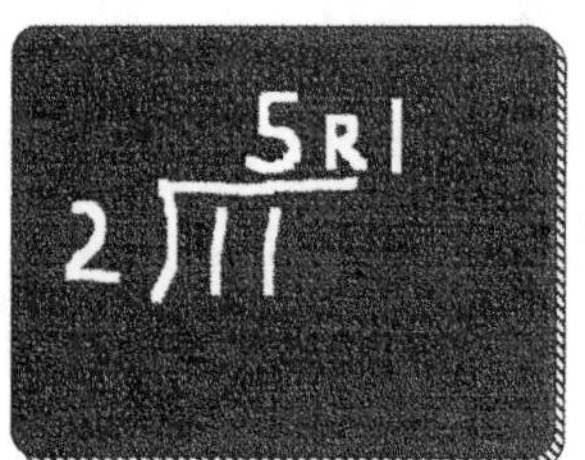

3–6 Interpreting Symbols

Whole-Class or Small-Group Activity

Materials: Counters ▪ Plastic or paper cups

Write division problems on the chalkboard and ask the children to act them out using the counters. Let the children choose how to divide the counters (by putting them into cups, snapping them together to form trains, or grouping them in piles). Then ask the children to tell how they modeled each problem. For example:

Write 12 ÷ 4.

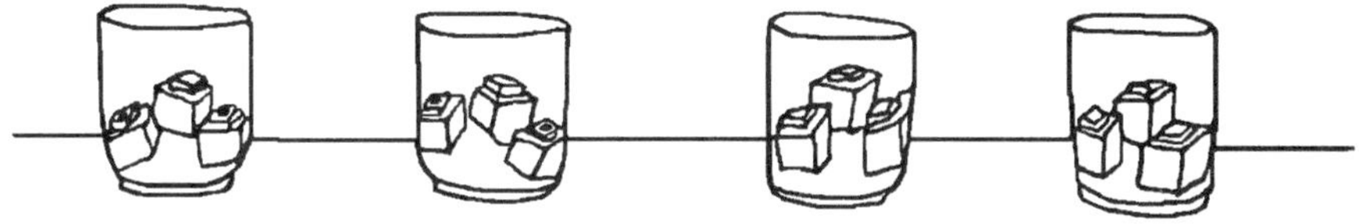

I used four cups. I put three into each cup.

I made four rows. I put three in each row.

Write 6 ÷ 2.

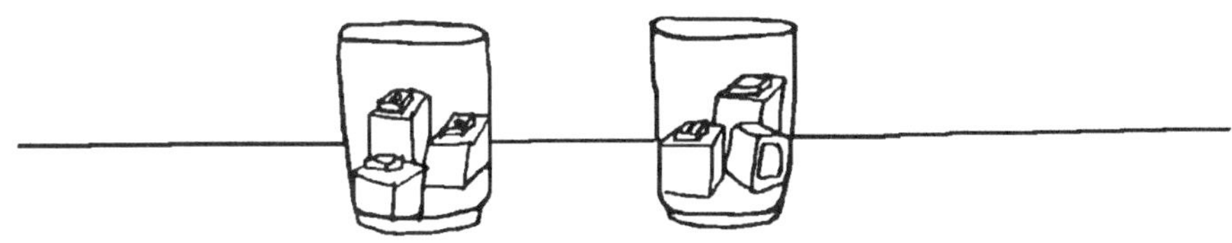

I made groups. Six cubes divided into two groups makes three in each group.

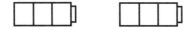

I broke up a train into two parts.

Variation: Have the children make up stories to go with the problems. Then have them draw or write their stories.

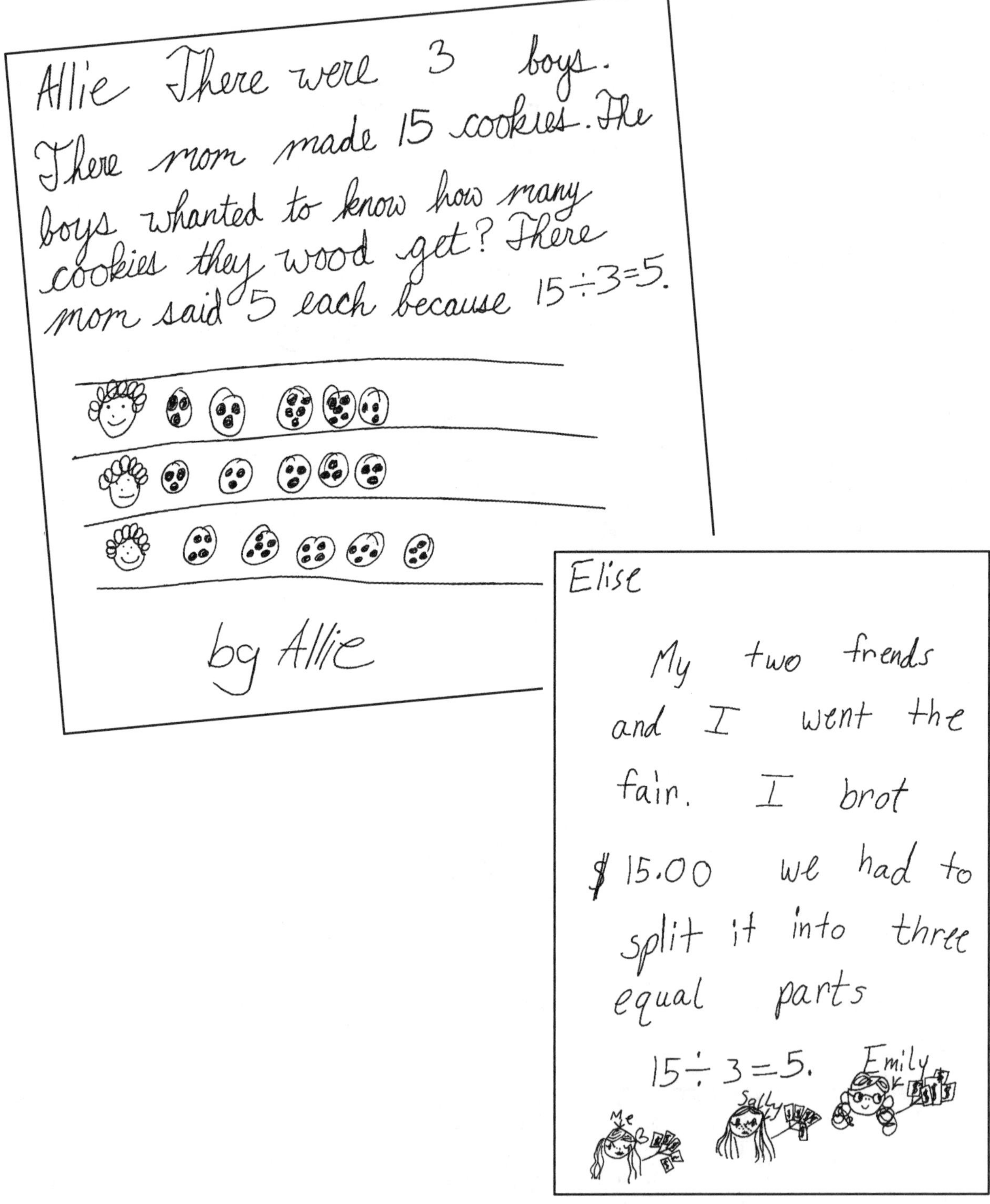

Allie There were 3 boys.
There mom made 15 cookies. The
boys whanted to know how many
cookies they wood get? There
mom said 5 each because 15÷3=5.
by Allie
Elise
My two frends
and I went the
fair. I brot
$15.00 we had to
split it into three
equal parts
15÷3=5.
Emily
Sally
Me

3–7 Learning To Write the Division Sign .... Whole-Class or Small-Group Activity

Materials: Connecting cubes, wooden cubes, Color Tiles, or collections ▪ Small chalkboard, chalk, and eraser for each child

After the children have had several experiences seeing you write division problems symbolically, they can begin writing them themselves.

Level 1: Copying Equations

Continue to tell story problems for the children to act out with counters. Write each corresponding equation using symbols, which the children can then copy onto their chalkboards.

Level 2: Writing Equations and Checking

After acting out a story problem with counters, have the children write the problem symbolically on their chalkboards. After they have finished, write the equation on the classroom chalkboard so that children can check what they wrote. Use both types of division symbols so that the children can see that either way is right.

3–8 Multiplication and Division Together: Story Problems .... Whole-Class or Small-Group Activity

Materials: Connecting cubes, wooden cubes, Color Tiles, or collections ▪ Small chalkboard, chalk, and eraser or pencil and paper for each child

When the children are comfortable with both multiplication and division, begin presenting opportunities for them to deal with both processes in the same activity.

As you randomly present both multiplication and division story problems, have the children use counters to represent the objects in the stories and write the equations that describe the actions. For each problem, have the children identify the appropriate process, multiplication or division. For example:

Sam had three packs of gum. Each pack had six pieces of gum in it. How many pieces of gum did Sam have? How did you figure out the answer? Did you multiply or divide?

Here's another one.

Nori's mom bought 24 treats for the party. There were ten children at the party. How many treats could each child get? How do you know? Did you multiply or divide?

Independent and Partner Activities

3–9 Counting Boards: Division

Independent Activity

Materials: *Level 1:* Connecting cubes (sorted by color), wooden cubes ▪ Counting boards (1 set of eight identical boards for each child) [BLMs #2–6] ▪ Division Cards [BLM #143]
Level 2: Same as for Level 1, plus 2" × 6" strips of paper (8 per child)
Level 3: Same as for Level 2, but without division cards

The children use the counting boards along with the division cards to practice working with division at a variety of levels.

Level 1: **Interpreting Division Problems**

Each child chooses a set of eight counting boards, a container of cubes, and eight division cards. The child spreads out the boards and puts a card below each. Then he or she places the cubes on the boards in groups to represent the problems shown on the cards.

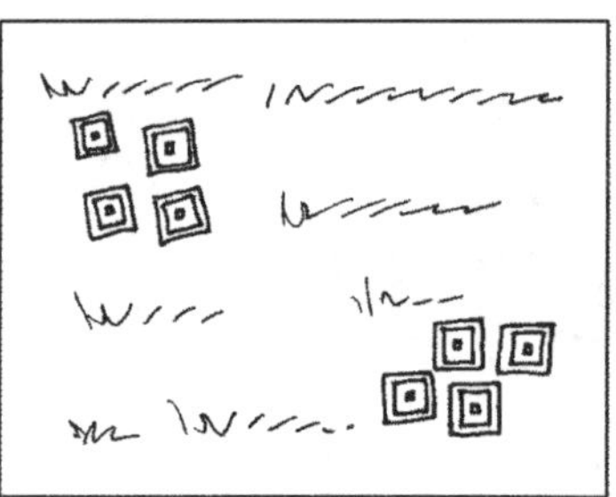

8 ÷ 2

Variation: The child may choose to use more than one counting board to represent a single division problem.

8 ÷ 2

Level 2: Copying and Solving Problems

The child proceeds as for Level 1 and then writes each problem and the answer on a 2" × 6" strip. After the child completes a series of problems, the papers can be stapled together to form a small book.

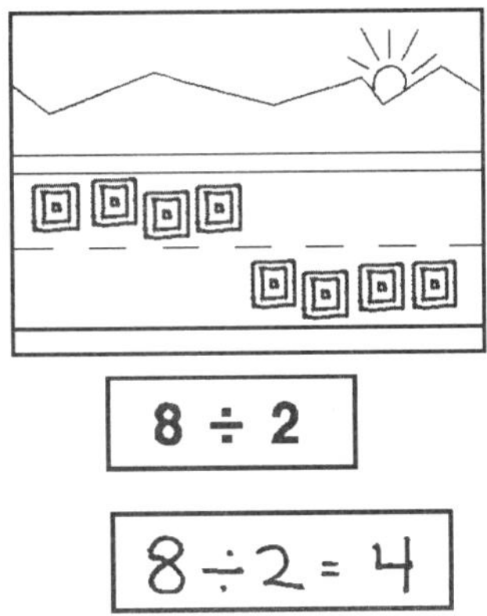

Variation: The child may choose to use more than one counting board to represent a single division problem.

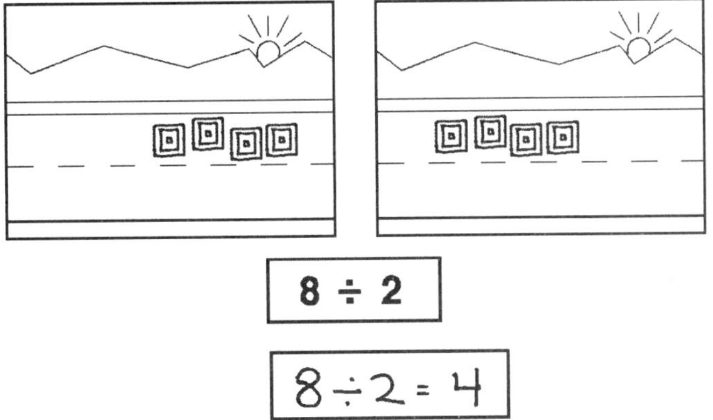

Level 3: Making Up and Solving Problems

Instead of using division cards to indicate problems, the child makes up his or her own problems, represents them on the counting boards, and writes them on 2" × 6" strips.

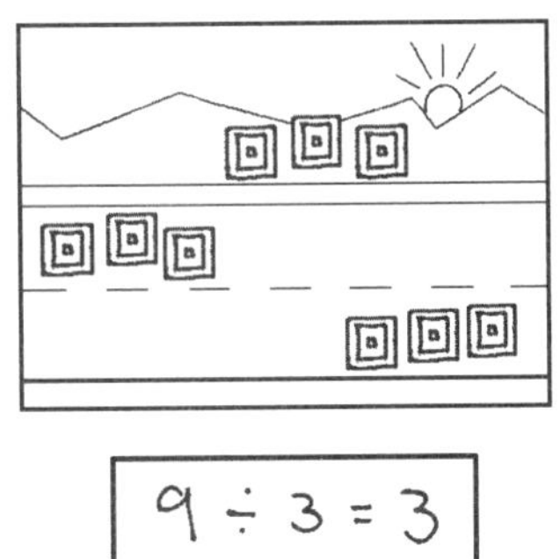

Variation: The child may choose to use several counting boards to represent a single division problem.

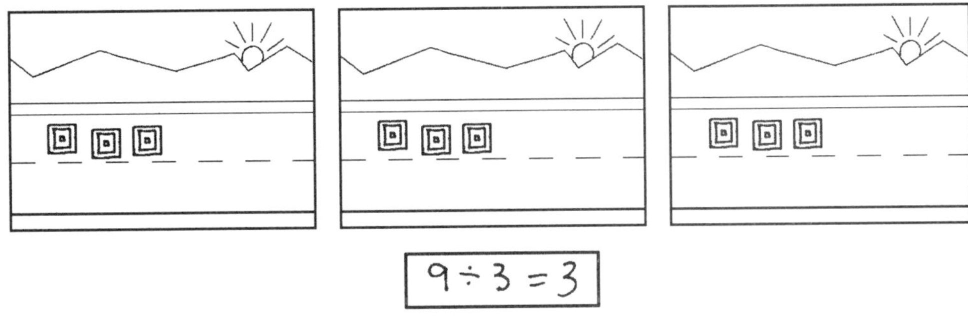

3–10 Number Shapes: Division

Independent Activity

Materials: Counters ▪ Number shapes (ten for each number) [BLMs #76–82] ▪ 2" × 6" strips of paper ▪ Division Cards [BLM #143] (optional)

The children choose a number shape and use it to help them divide quantities by the number it represents. They record the corresponding equation on a 2" × 6" strip. Children may determine the number of counters in one of several ways—by picking the numbers themselves, taking a handful of counters, or by using the Division Cards. For example:

I am going to work with the "seven" number shape. I will divide 26 by seven. First I'll get 26 cubes. I'll use them to fill as many of these number shapes as I can.

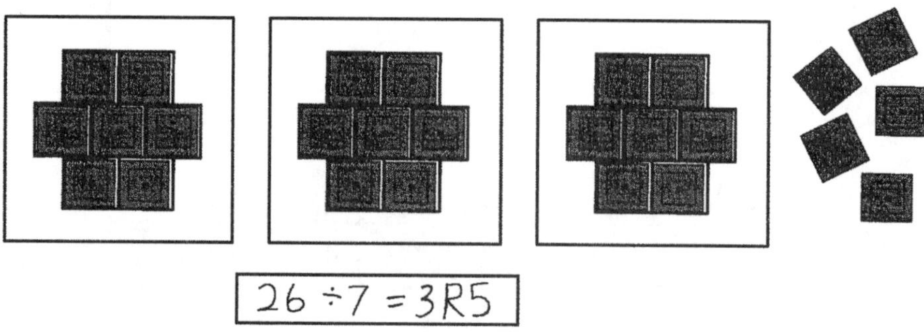

I have five left over. That's not enough to fill another "seven" number shape.

Now I am going to divide nineteen by seven.

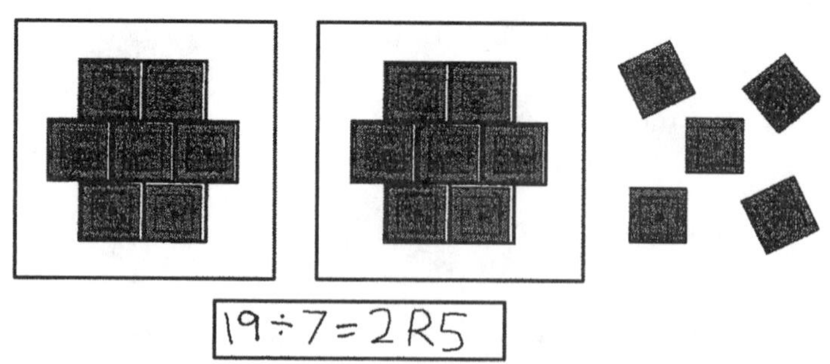

I started with 19 tiles. I filled two "seven" shapes and I have five left over.

3–11 Making Rows .... Independent Activity

Materials: Connecting cubes or Color Tiles ▪ Making Rows Worksheets [BLM #141] ▪ Making Rows Cards (See preparation below.) ▪ 1–6 Number cubes (1 per child)
Preparation: Duplicate BLM #141 and cut each sheet into three strips of four problems each. Note that copies of these worksheet strips can be used repeatedly because new problems will be created each time the children roll the number cubes. Make the Making Rows Cards by drawing large division symbols on 9" × 12" pieces of tagboard, as shown below.

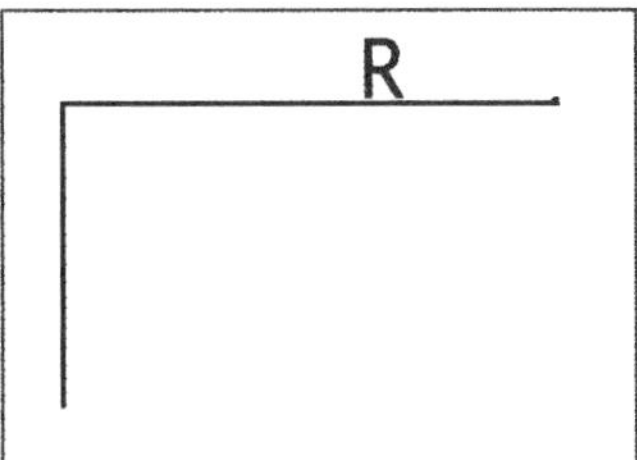

In this activity, the children create their own division problems using the division symbol (Making Rows Cards), counters, worksheet strips, and number cubes. Be sure to model this activity before expecting the children to work independently.

The child considers the first problem on a worksheet strip and then rolls a number cube to determine the divisor and writes it in place on the strip. The child models this division situation with counters on the Making Rows Card and then writes the answer on the strip. A child may complete as many division problems as time allows. For example:

I rolled a three. It tells me how many rows to make on the card.

The child writes “3” for the divisor in the first problem on the worksheet strip.

Now I’ll get twenty cubes and put them in three rows on the card.

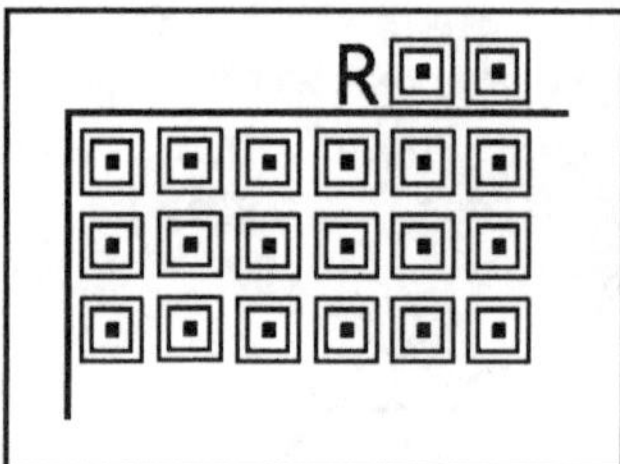

Oh, two are left over. They go on top. That is the remainder.

There are six in each row and two left over. I’ll write that down on my worksheet strip.

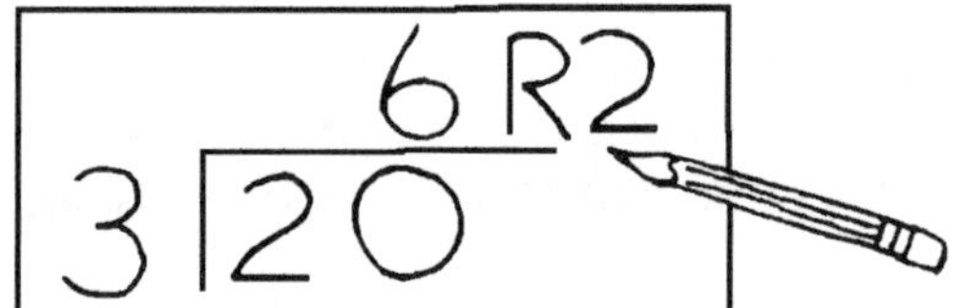

3–12 Problems for Partners: Division .... Independent Activity for Partners

Materials: Connecting cubes ▪ Small chalkboard, chalk, and eraser or pencil and paper for each child

Partners make up division problems together, taking turns picking the numbers for the dividend and the divisor. For example:

The first partner picks the number of cubes to be divided and makes a train of that length.

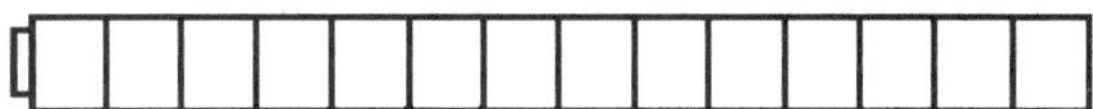

This train is fourteen cubes long.

The second partner then decides how to break up the train into equal groups.

Let's break up the train into twos.

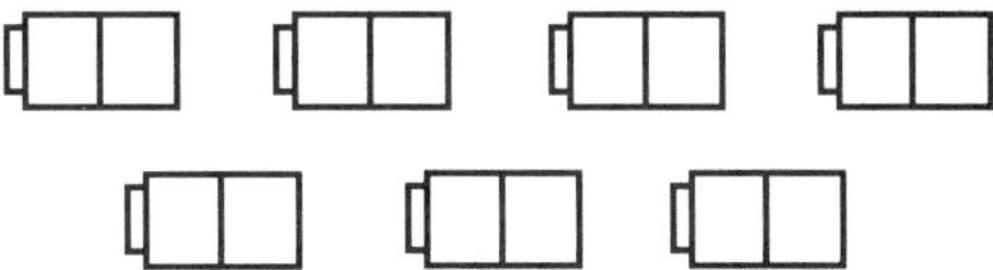

The first partner writes the equation that describes how the second partner broke up the train and the number of groups that resulted.

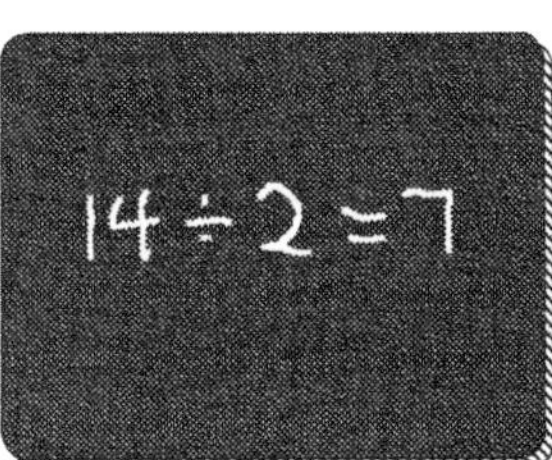

The partners continue to make up, model, and record division problems, taking turns for each problem.

3–13 Cups of Cubes

Independent Activity

Materials: Connecting cubes • Plastic cups (at least nine) • Dividing Strips Worksheets [BLM #142]

The child chooses the number of cups with which to work and then considers a problem on a Dividing Strip to determine how many cubes to divide among the chosen number of cups. After modeling the problem, the child completes the equation on the strip. For example:

I will work with three cups and sixteen cubes.

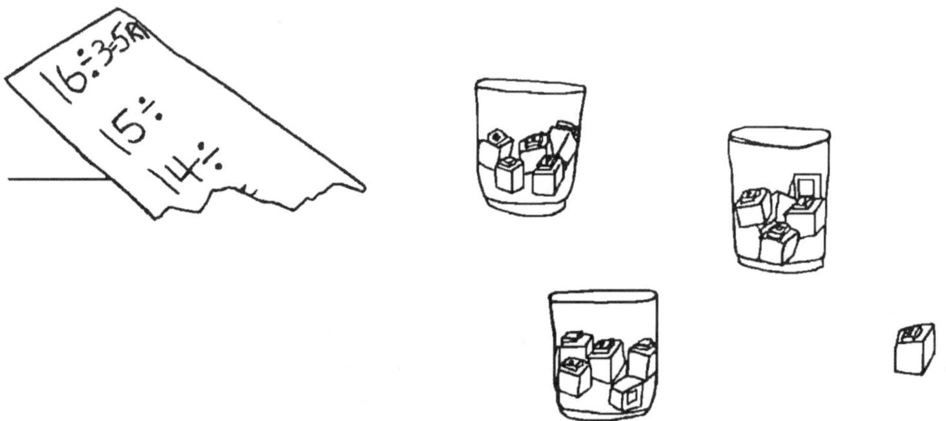

That's five in each cup and one left over.

3–14 How Many Buildings? .... Independent Activity

Materials: Connecting cubes • Dividing Strips Worksheets [BLM #142] • 1–6 Number cubes (1 per child)

In this activity, the children find out how many buildings of a certain height they can make from a given number of cubes. The child considers a problem on a Dividing Strip to determine a total number of cubes. Then the child rolls a number cube to determine how high to make each building and records that number on the strip.

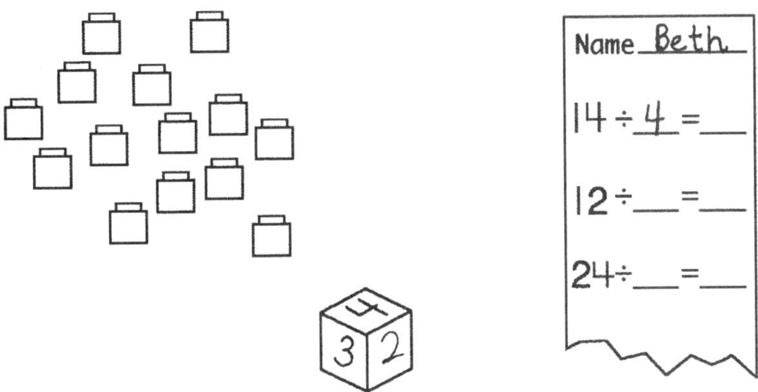

I started with 14 cubes. I rolled a four, so I must make each building four cubes high.

After modeling the buildings, the child records the results on the strip.

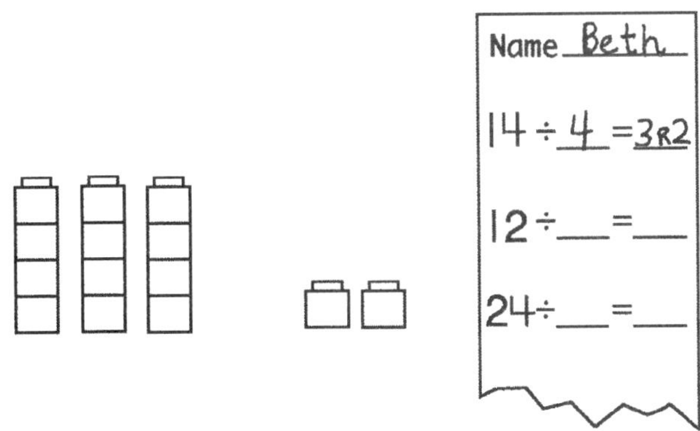

I can make three buildings with two cubes left over.

3–15 Creation Cards for Division .... Independent Activity

Materials: Connecting cubes • Creation Cards [BLMs #7–12] • Dividing Strips Worksheets [BLM #142]

Each child chooses a Creation Card to determine a divisor. The task is to determine how many creations of this size can be made from each of the various numbers of cubes indicated on the Dividing Strips. The child records the answers on the strips. For example:

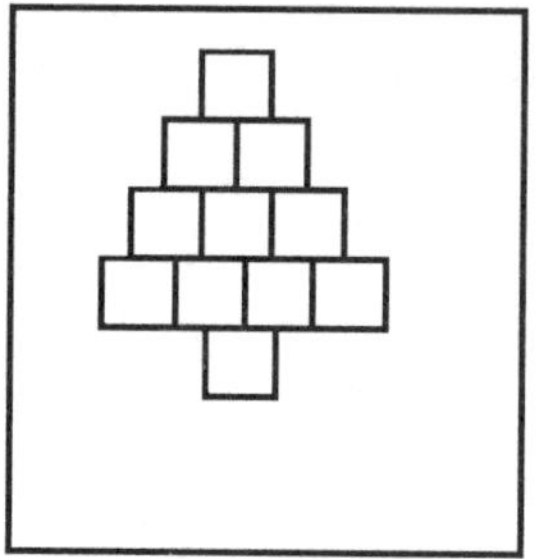

I picked a tree. It takes eleven cubes.

Name Caitlin

25 ÷ 11 = ___
32 ÷ ___ = ___
12 ÷ ___ = ___
28 ÷ ___ = ___

My first number is 25. How many trees can I make?

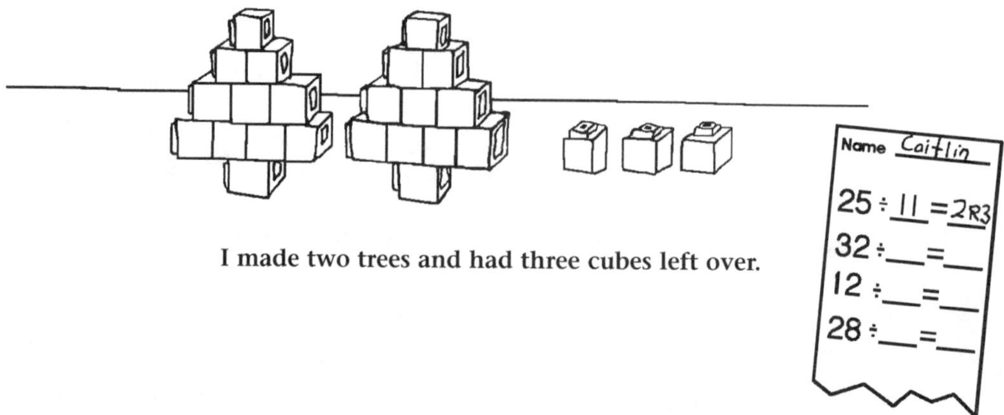

Name Caitlin
25 ÷ 11 = 2R3
32 ÷ ___ = ___
12 ÷ ___ = ___
28 ÷ ___ = ___

I made two trees and had three cubes left over.

Blackline Masters

Here is a complete list of the blackline masters included in the *Developing Number Concepts* series. The masters that are to be used with *Book Three* are listed in heavy type.

1 Working-Space Paper
2 Counting Boards (Tree/Ocean)
3 Counting Boards (Barn/Cave)
4 Counting Boards (Corral/Store)
5 Counting Boards (Road/House)
6 Counting Boards (Garden/Grass)
7 Creation Cards (Doorway/Pig)
8 Creation Cards (Tree/Caterpillar)
9 Creation Cards (Horse/Giraffe)
10 Creation Cards (Dog/Table)
11 Creation Cards (Slide/Fireplace)
12 Creation Cards (Robot/Bench)
13 Small Dot Cards (1–10 dots)
14 Numeral Cards 0–6 (front)
15 Numeral Cards 0–6 (back)
16 Numeral Cards 0–10 (front)
17 Numeral Cards 0–10 (back)
18 Numeral Cards 11–20 (front)
19 Numeral Cards 11–20 (back)
20 Tell-Me-Fast Dot Cards
21 Tell-Me-Fast Dot Cards
22 Tell-Me-Fast Dot Cards
23 Tell-Me-Fast Dot Cards
24 Tell-Me-Fast Dot Cards
25 Tell-Me-Fast Dot Cards
26 Tell-Me-Fast Dot Cards
27 Tell-Me-Fast Dot Cards
28 Number Lines (1–10)
29 Number Lines (1–20)
30 Roll-a-Tower Game Board (1–6)
31 Roll-a-Tower Game Board (4–9)
32 Build a City/Building Stacks Game Board
33 Counting Worksheet
34 Shape Puzzles (3–6)
35 Shape Puzzles (7–10)
36 Shape Puzzles (7–10)
37 Shape Puzzles (10–20)
38 Shape Puzzles (10–20)
39 Shape Puzzles (10–20)
40 Shape Puzzles (10–20)
41 Line Puzzles (3–6)
42 Line Puzzles (3–6)
43 Line Puzzles (3–6)
44 Line Puzzles (3–6)
45 Line Puzzles (7–10)
46 Line Puzzles (7–10)
47 Line Puzzles (7–10)
48 Line Puzzles (10–20)
49 Line Puzzles (10–20)
50 Line Puzzles (10–20)
51 Line Puzzles (10–20)
52 Line Puzzles (10–20)
53 Line Puzzles (10–20)
54 Line Puzzles (10–20)
55 Line Puzzles (10–20)
56 Hands Worksheet
57 Measure-It Worksheet
58 Fill-It Worksheet
59 Colors Worksheet
60 Pattern-Train Outlines
61 Rhythmic-Motions Pictures
62 More/Less/Same Cards
63 More/Less/Same Worksheet
64 Addition Cards (Sums to 6)
65 Addition Cards (Sums to 6)
66 Addition Cards (Sums of 7 to 9)
67 Addition Cards (Sums of 7 to 9)
68 Addition Cards (Sums of 10)
69 Subtraction Cards (Subtracting from 1 to 6)
70 Subtraction Cards (Subtracting from 1 to 6)
71 Subtraction Cards (Subtracting from 7 to 9)
72 Subtraction Cards (Subtracting from 7 to 9)
73 Subtraction Cards (Subtracting from 10)
74 Clear-the-Deck Game Board
75 Hiding Assessment Recording Sheet
76 Number Shapes (4)
77 Number Shapes (5)
78 Number Shapes (6)
79 Number Shapes (7)
80 Number Shapes (8)
81 Number Shapes (9)
82 Number Shapes (10)

83 Number Shapes Recording Sheet (4)
84 Number Shapes Recording Sheet (5)
85 Number Shapes Recording Sheet (6)
86 Number Shapes Recording Sheet (7)
87 Number Shapes Recording Sheet (8)
88 Number Shapes Recording Sheet (9)
89 Number Shapes Recording Sheet (10)
90 Number Shapes Equations Worksheet
91 Number Train Outlines (3)
92 Number Train Outlines (4)
93 Number Train Outlines (5)
94 Number Train Outlines (6)
95 Number Train Outlines (7)
96 Number Train Outlines (8)
97 Number Train Outlines (9)
98 Number Train Outlines (10)
99 Number Trains Recording Sheet (3)
100 Number Trains Recording Sheet (4)
101 Number Trains Recording Sheet (5)
102 Number Trains Recording Sheet (6)
103 Number Trains Recording Sheet (7)
104 Number Trains Recording Sheet (8)
105 Number Trains Recording Sheet (9)
106 Number Trains Recording Sheet (10)
107 Number Trains Equations Worksheet
108 Equations Worksheet
109 Build-a-Floor Game Board
110 Making Numbers Worksheet (3 to 7)
111 Making Numbers Worksheet (8 to 10)
112 Apartment Buildings Game Boards
113 Place-Value Strips
114 Graph Paper
115 10 × 10 Matrix
116 00–99 Chart
117 00–99 Charts Recording Sheet
118 Grid Picture Task Card A
119 Grid Picture Task Card B
120 Grid Picture Task Card C
121 Grid Picture Task Card D
122 Grid Picture Task Card E
123 Grid Picture Task Card F
124 Grid Picture Task Card G
125 Grid Picture Task Card H
126 Tens and Ones Worksheet
127 Hundreds, Tens, and Ones Worksheet
128 How Many More? Worksheet
129 Measuring Things Worksheet
130 Compared-to-Me Worksheet
131 Measuring Myself Worksheet
132 Making Trails Worksheet
133 Roll-and-Add Worksheet
134 Roll-and-Subtract Worksheet
135 How Many Groups? Worksheet
136 Multiplying Strips Worksheet
137 Multiplication Cards
138 Multiplication Cards
139 Patterns in Multiplying Worksheet (Constant Groups)
140 Patterns in Multiplying Worksheet (Sequential Groups)
141 Making Rows Worksheet
142 Dividing Strips Worksheet
143 Division Cards

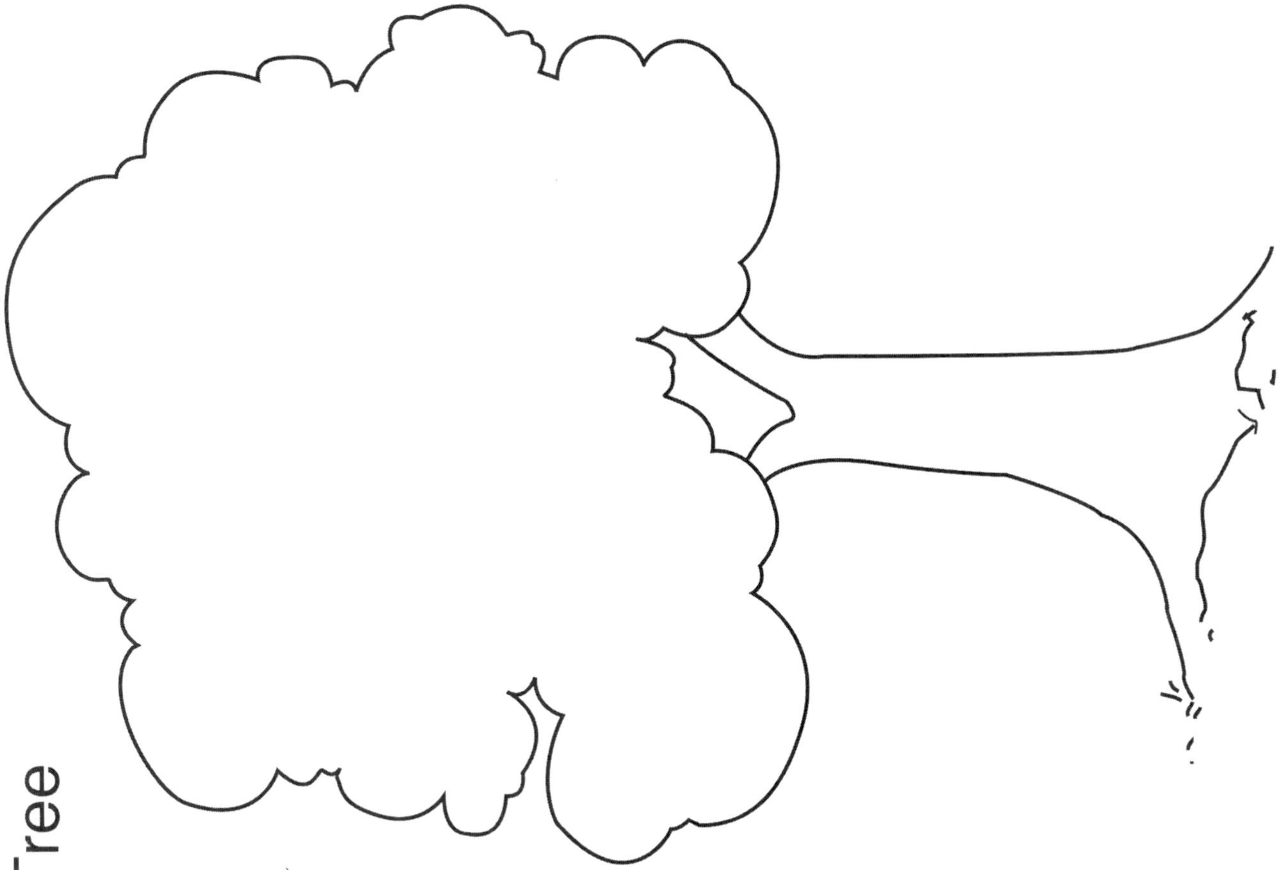

Tree

cut

Ocean

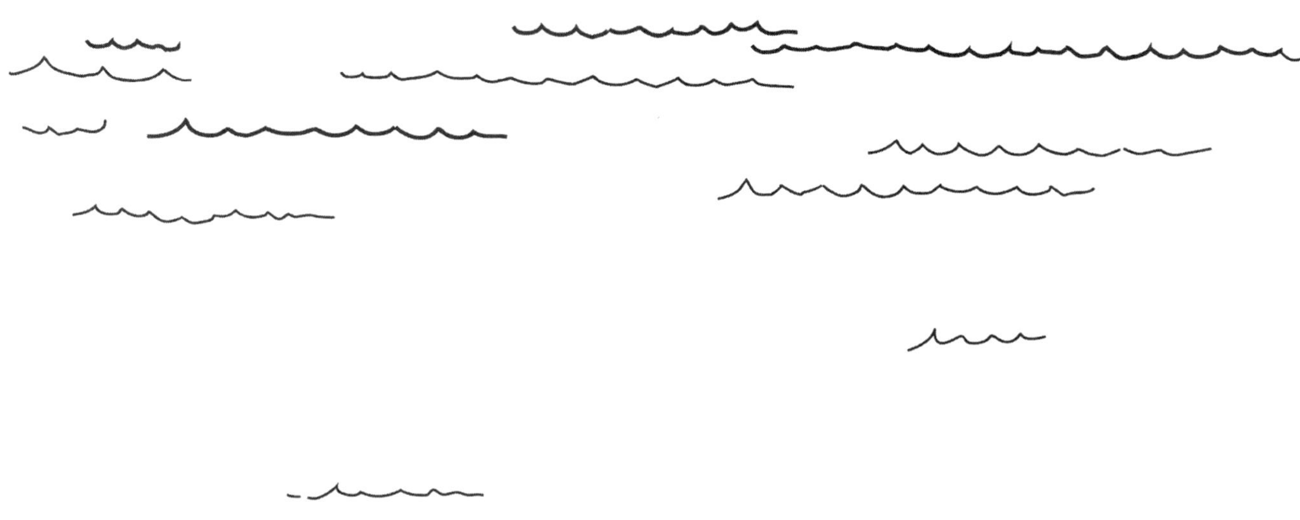

© by Math Perspectives

Barn

cut

Cave

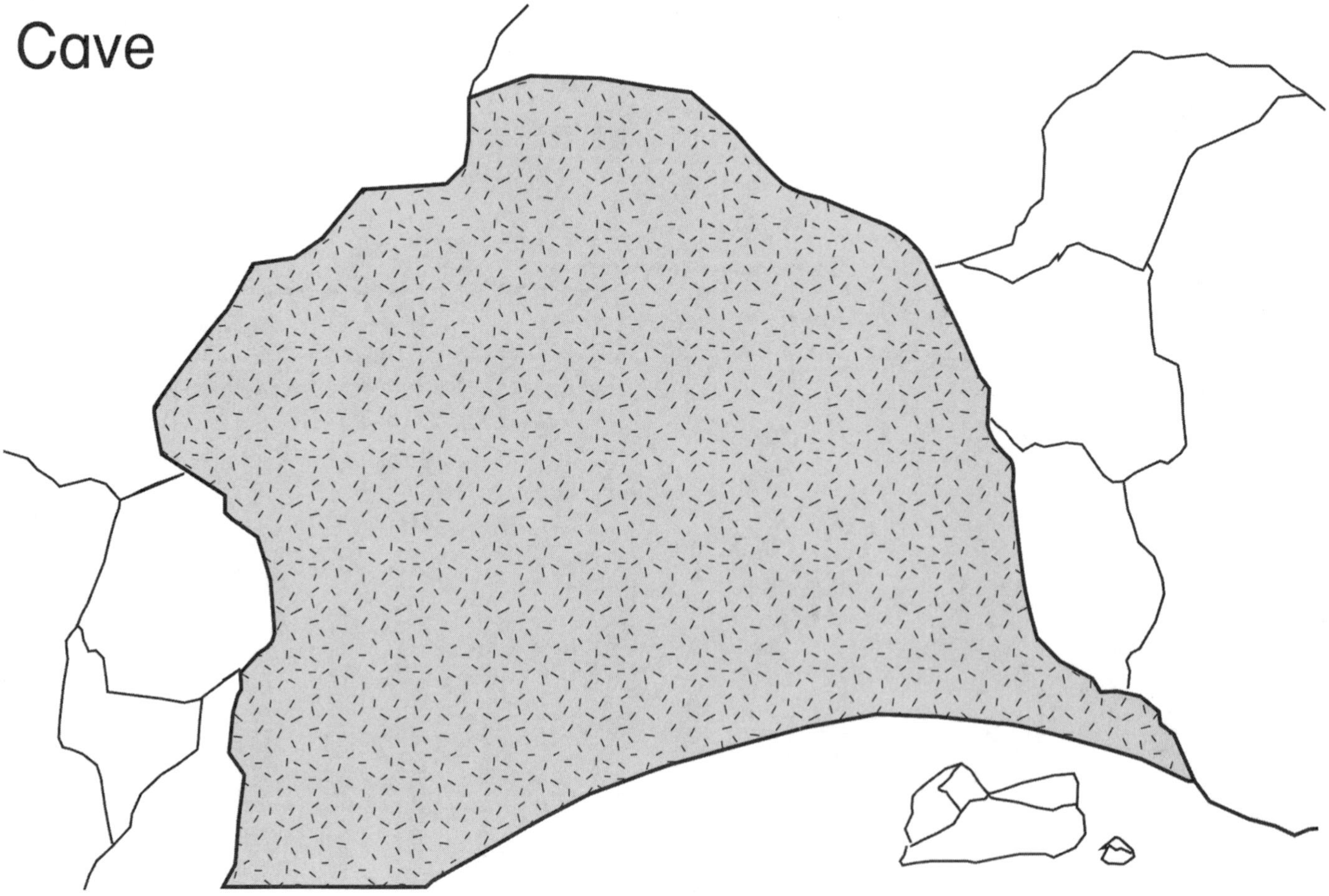

© by Math Perspectives

Corral

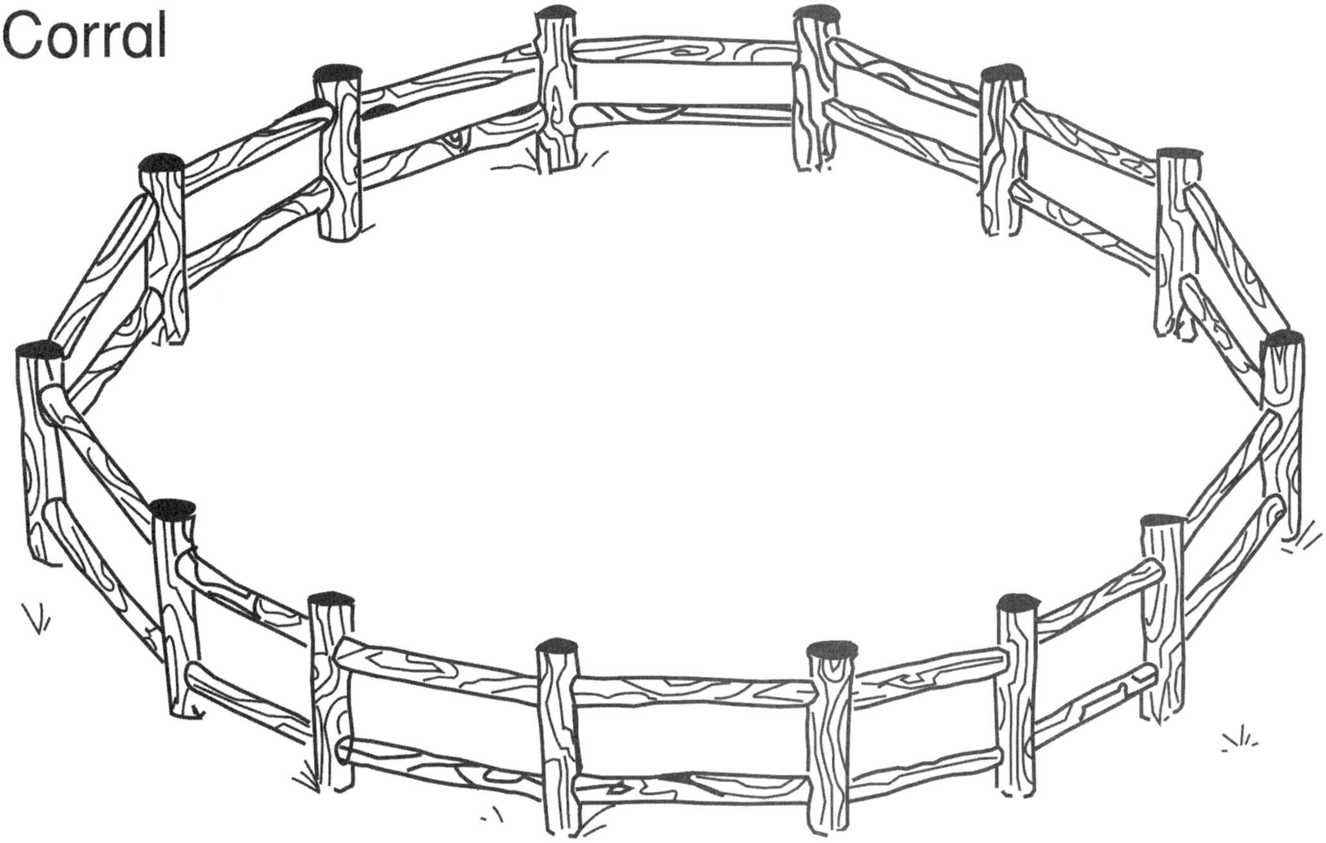

cut

Store

STORE

© by Math Perspectives

Road

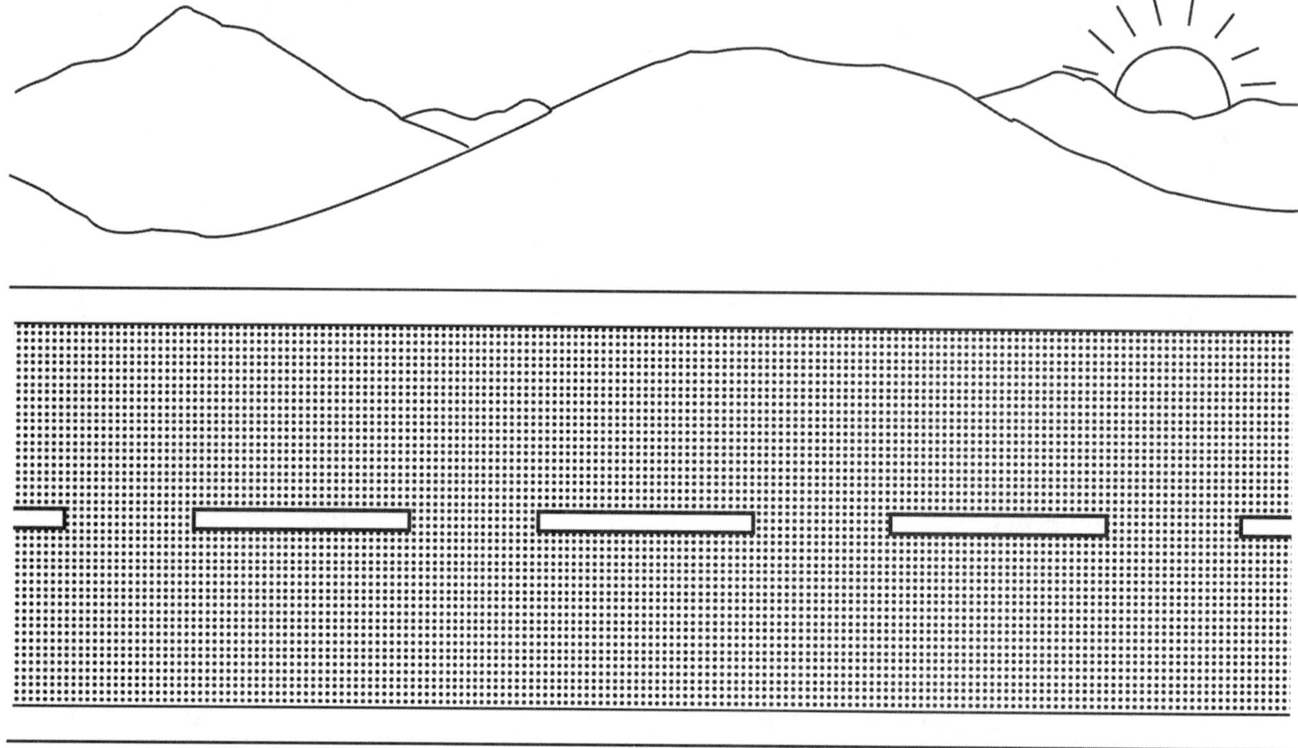

cut

House

© by Math Perspectives

Garden

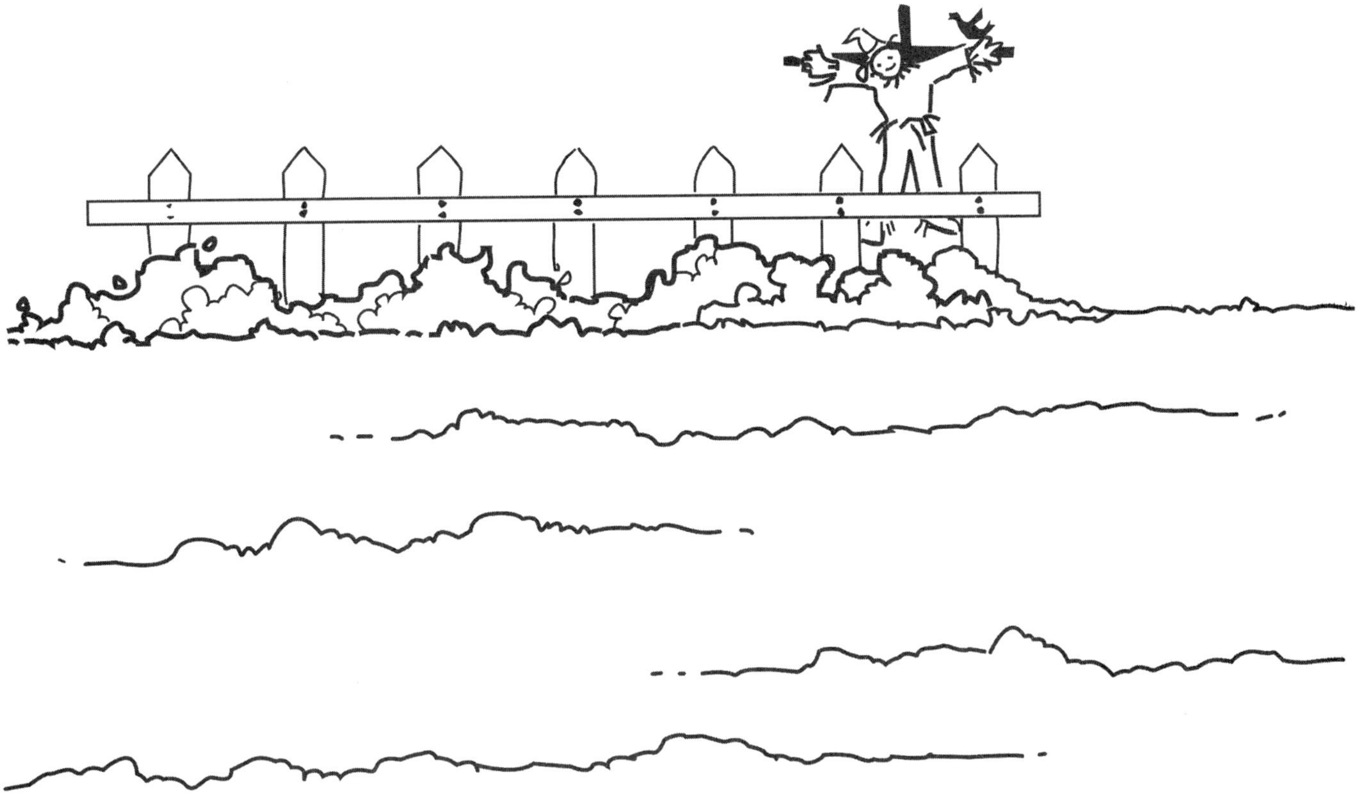

cut

Grass

© by Math Perspectives

Doorway

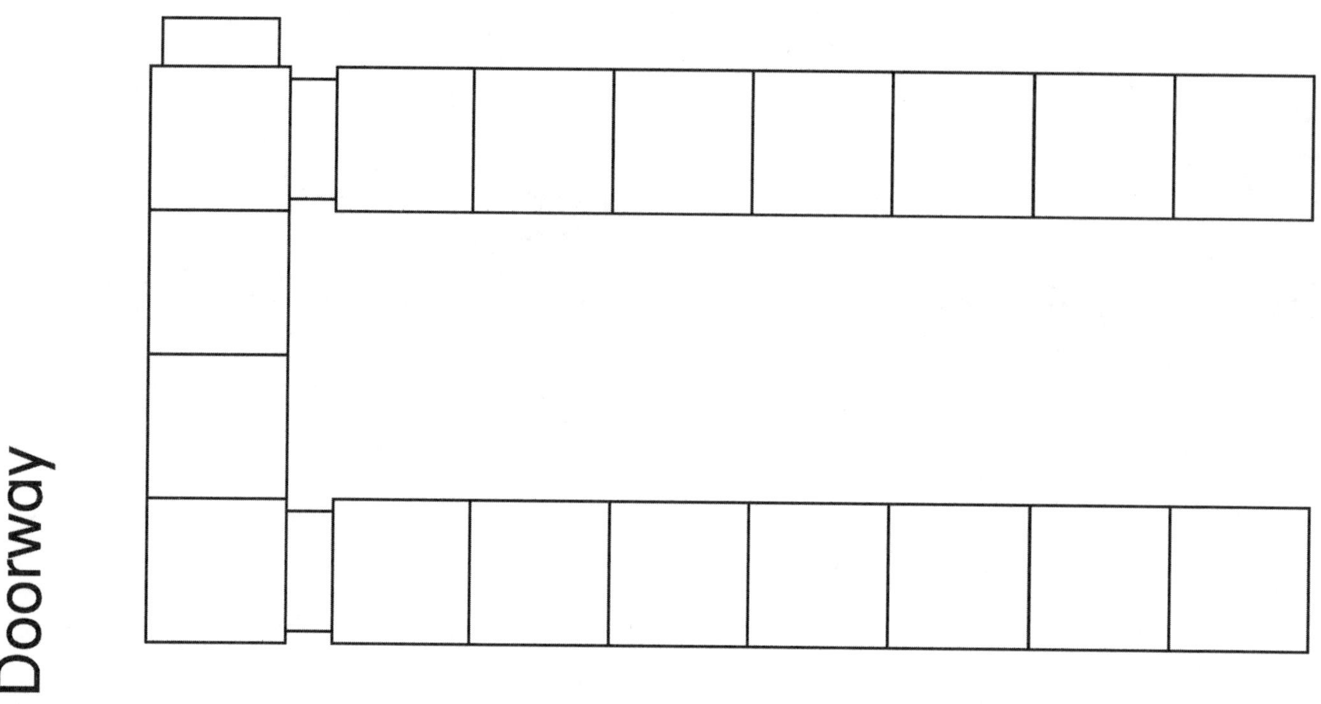

cut

Pig

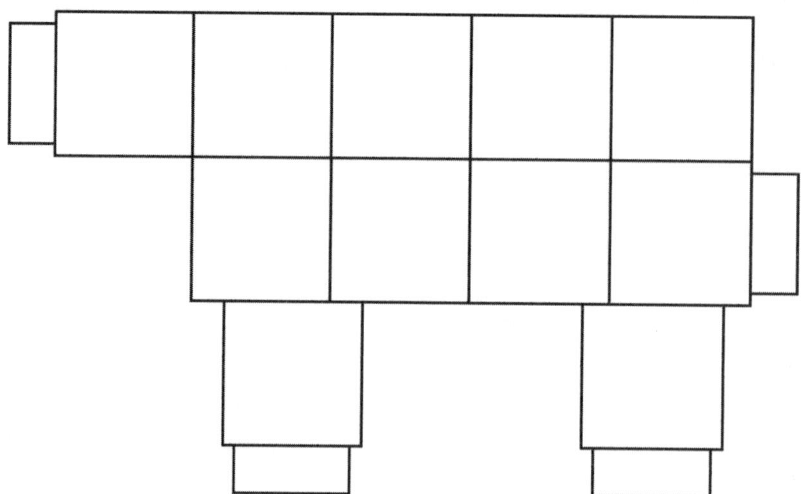

© by Math Perspectives

Caterpillar

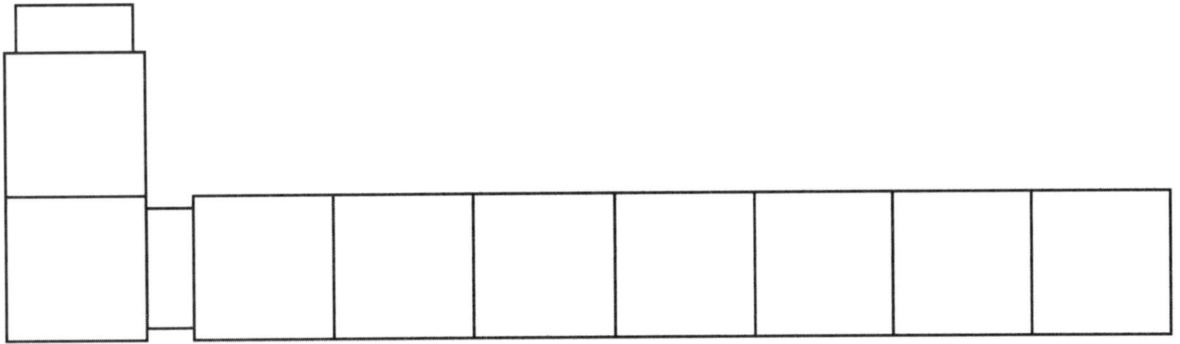

cut

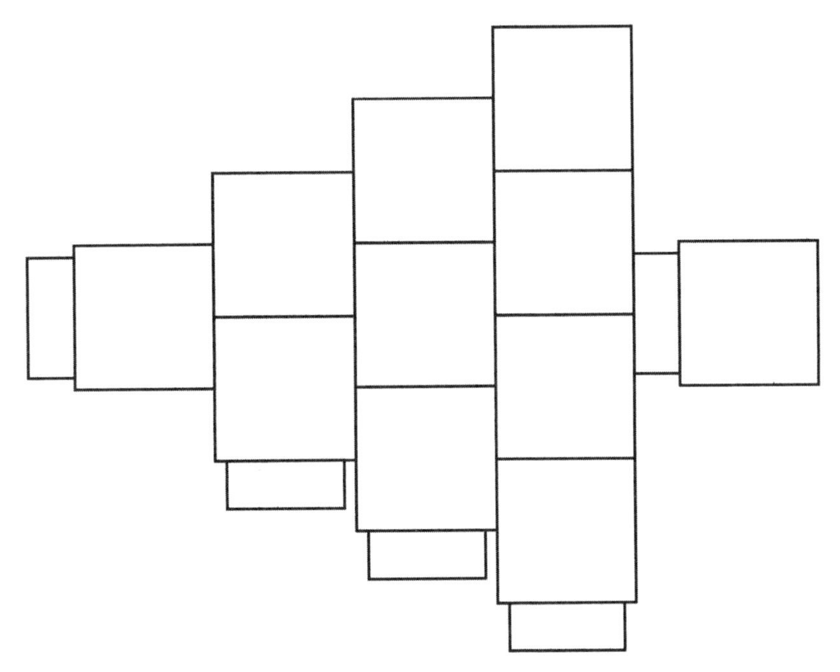

Tree

© by Math Perspectives

Horse

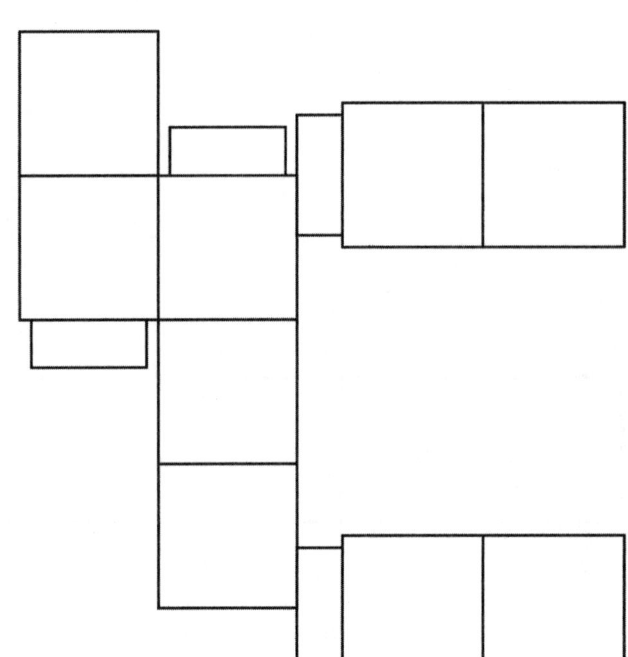

cut

Giraffe

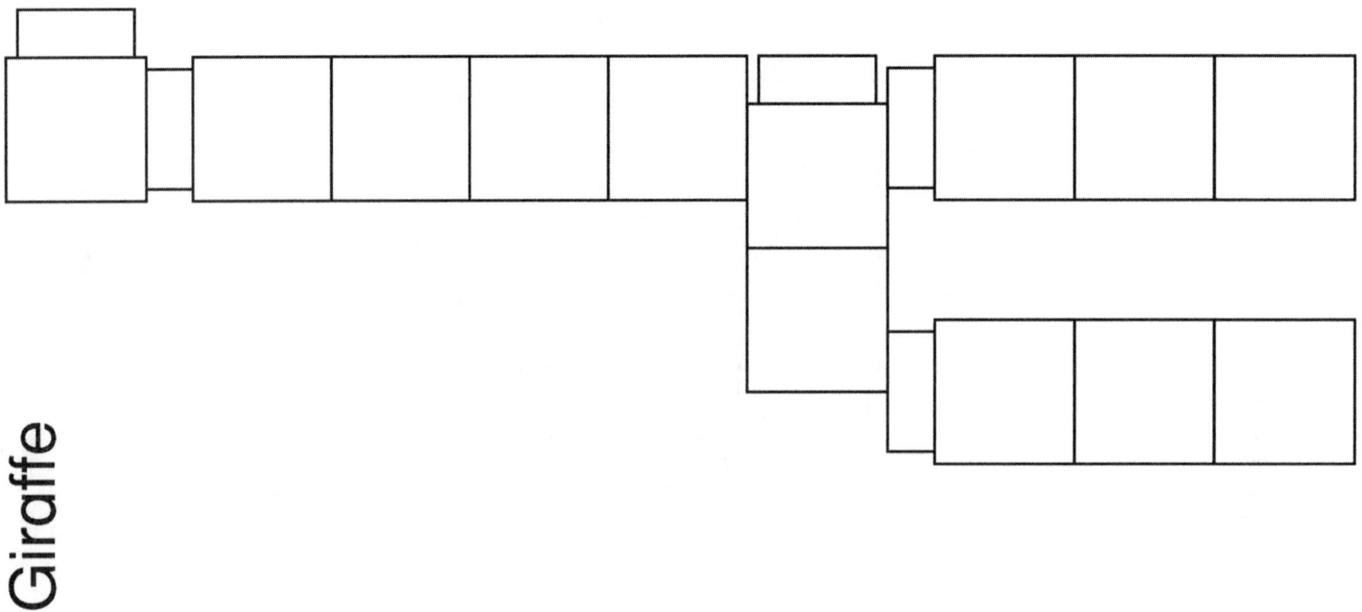

© by Math Perspectives

Table

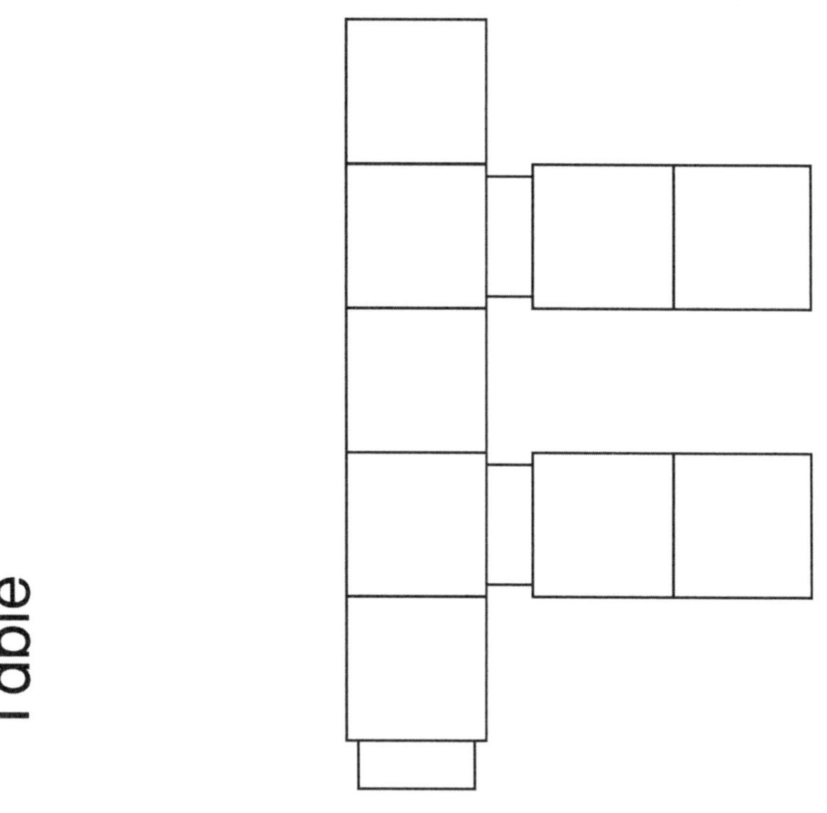

cut

Dog

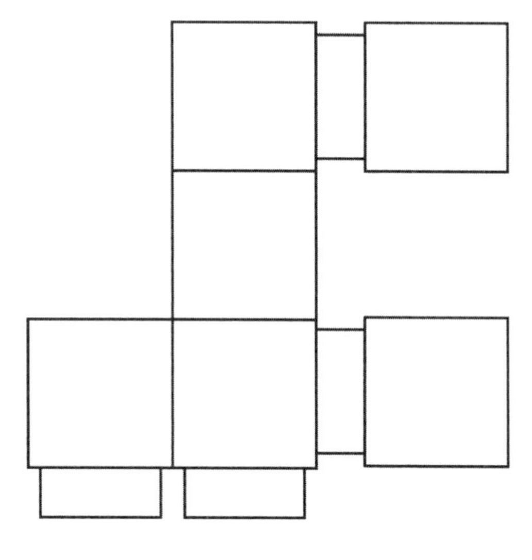

© by Math Perspectives

Fireplace

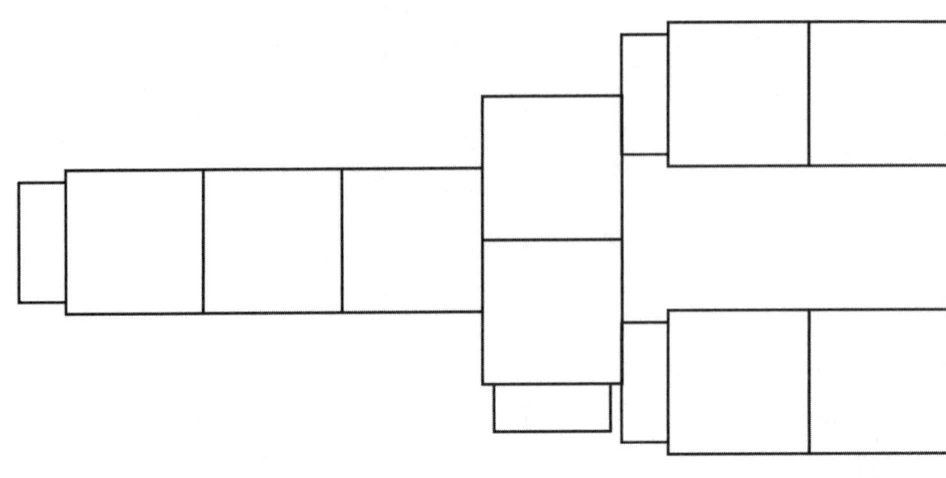

cut

Slide

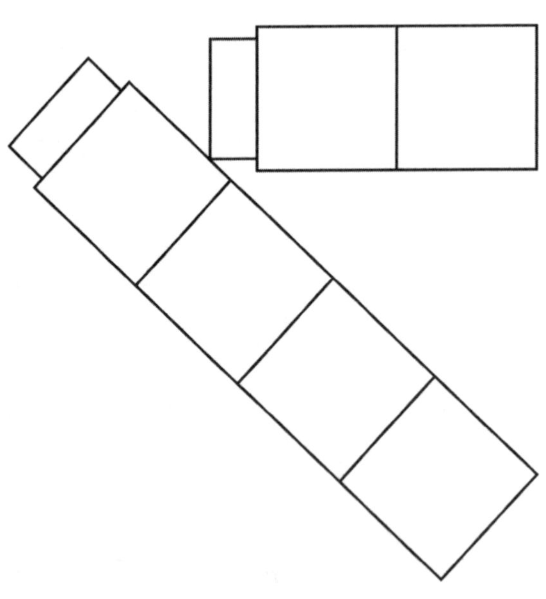

© by Math Perspectives

Robot

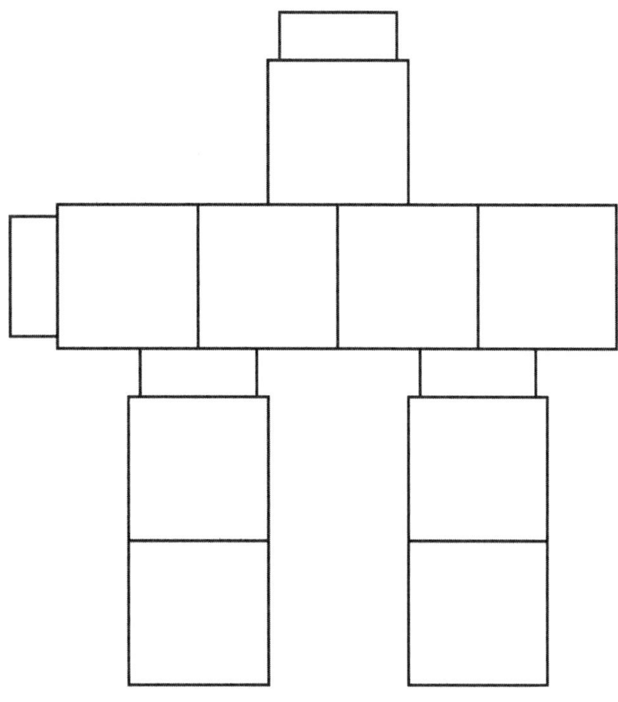

cut

Bench

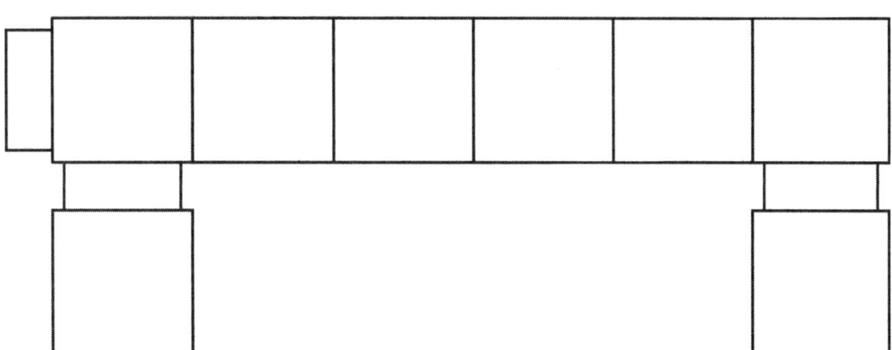

© by Math Perspectives

cut

© by Math Perspectives

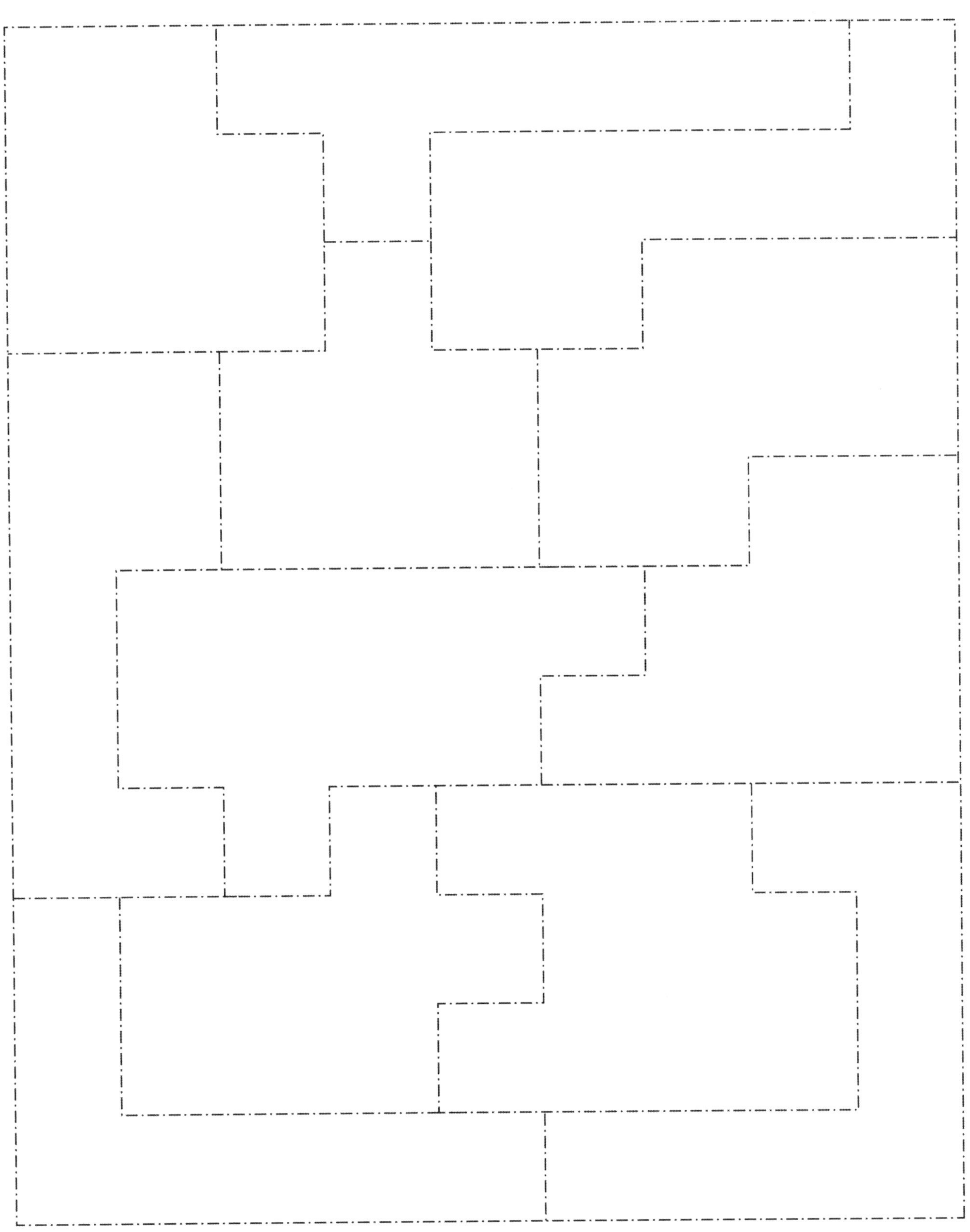

© by Math Perspectives

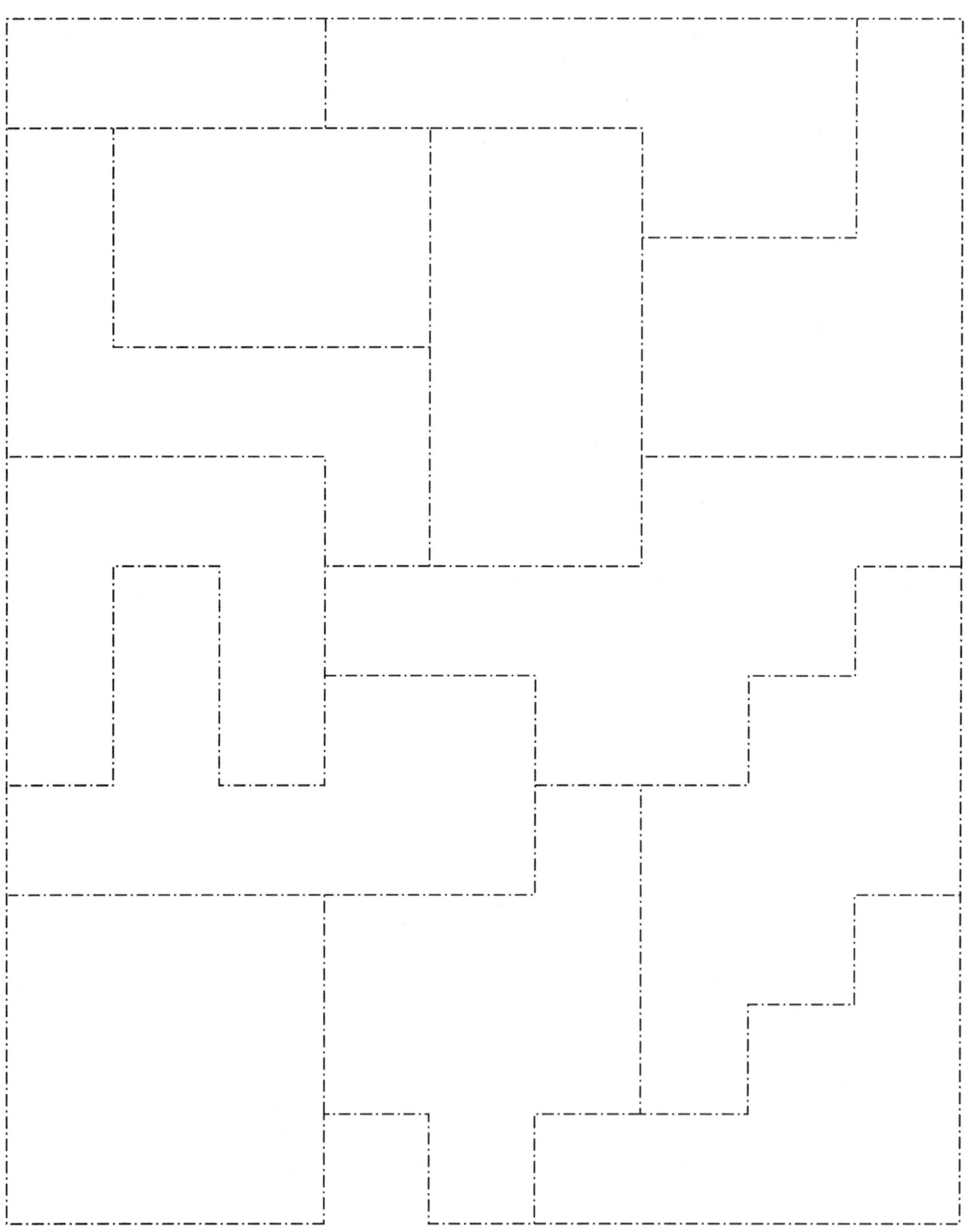

© by Math Perspectives

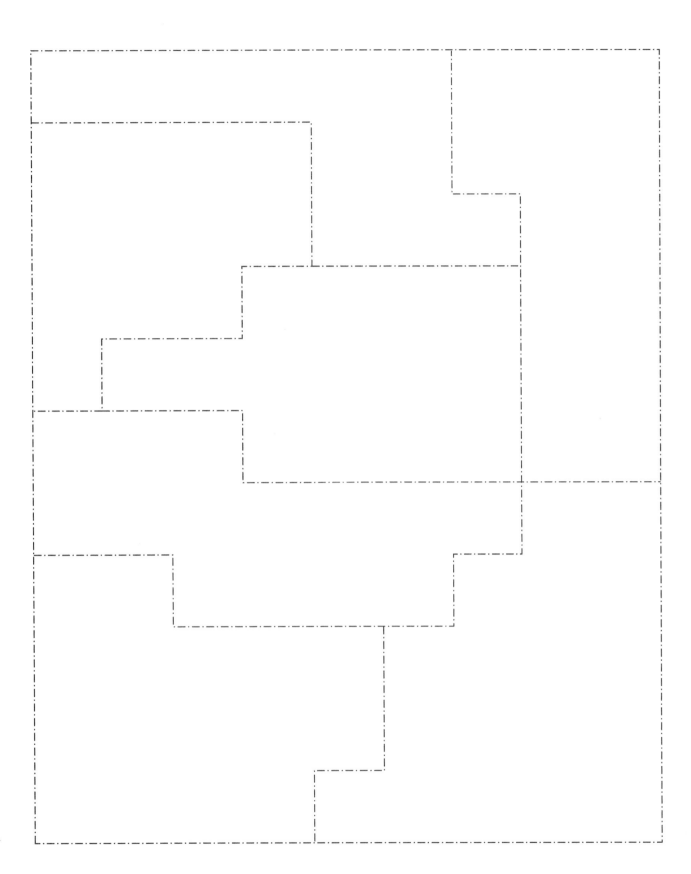

© by Math Perspectives

© by Math Perspectives

© by Math Perspectives

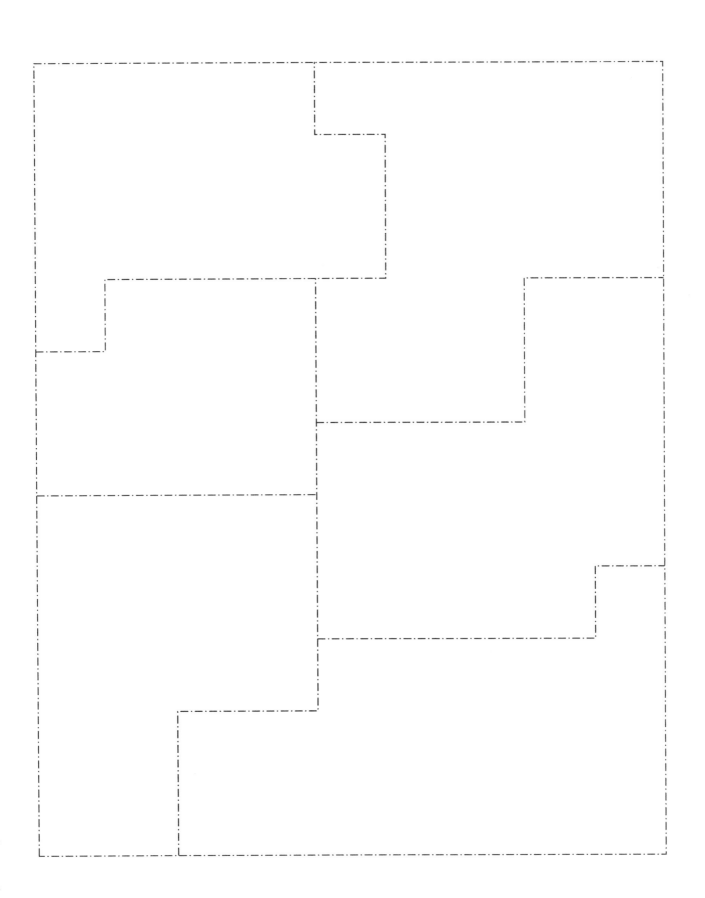

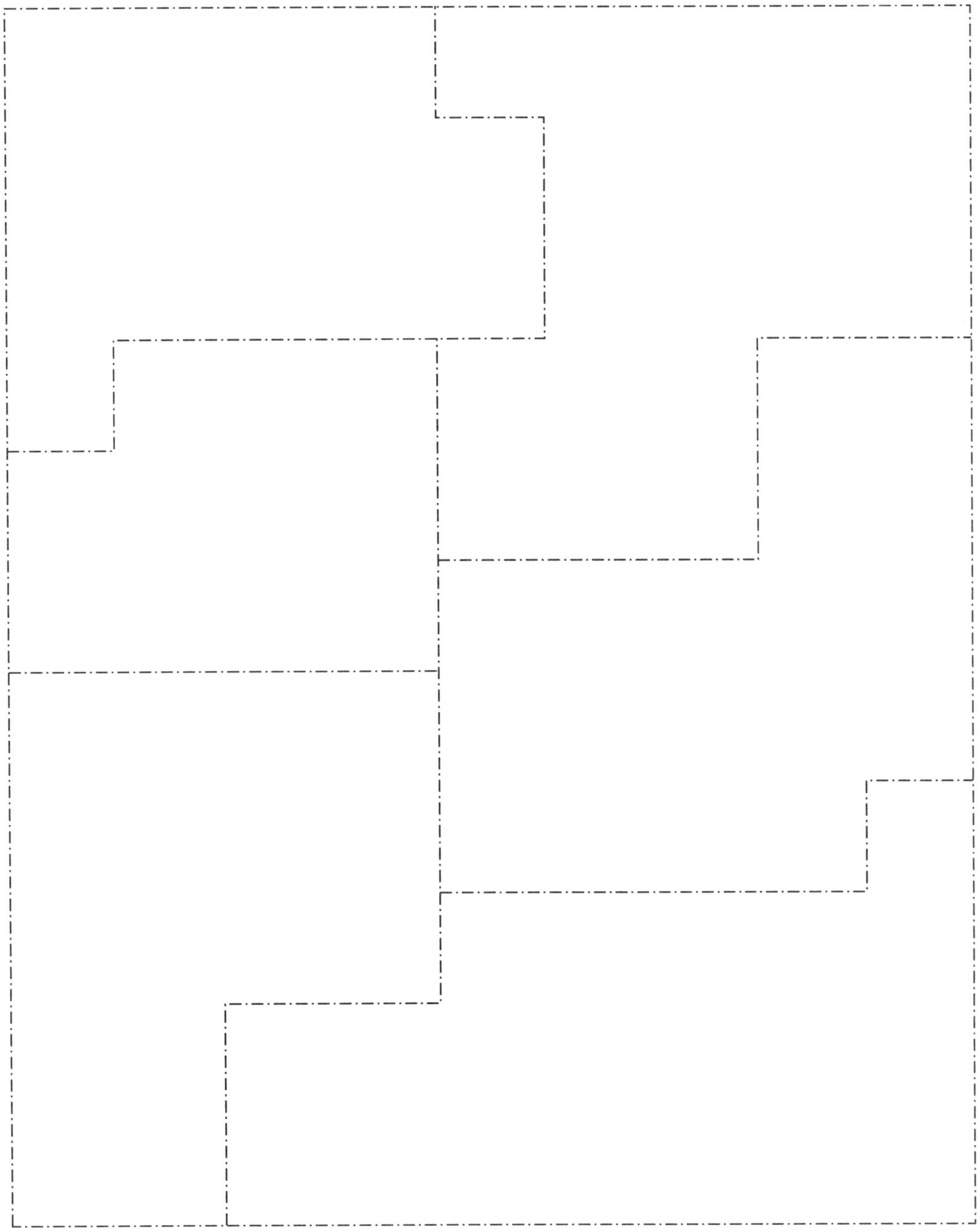

Name

Is it more, less, or the same?

> < =

© by Math Perspectives

cut

1 + 0	4 + 0	1 + 4
0 + 1	3 + 1	0 + 5
2 + 0	2 + 2	6 + 0
1 + 1	1 + 2	5 + 1
0 + 2	0 + 4	4 + 2
3 + 0	5 + 0	3 + 3
2 + 1	4 + 1	2 + 4
1 + 2	3 + 2	1 + 5
0 + 3	2 + 3	0 + 6

cut

cut

cut

© by Math Perspectives

cut

$1 + 0$	$0 + 1$	$2 + 0$	$1 + 1$	$0 + 2$	$3 + 0$	$0 + 3$	$1 + 2$	$2 + 1$
$4 + 0$	$0 + 4$	$1 + 3$	$3 + 1$	$2 + 2$	$5 + 0$	$4 + 1$	$3 + 2$	$2 + 3$
$1 + 4$	$0 + 5$	$6 + 0$	$1 + 5$	$2 + 4$	$3 + 3$	$4 + 2$	$5 + 1$	$0 + 6$

cut

© by Math Perspectives

cut

7 + 0	7 + 1	8 + 1
6 + 1	6 + 2	7 + 2
5 + 2	5 + 3	6 + 3
4 + 3	4 + 4	5 + 4
3 + 4	3 + 5	4 + 5
2 + 5	2 + 6	3 + 6
1 + 6	1 + 7	2 + 7
0 + 7	0 + 8	1 + 8
8 + 0	9 + 0	0 + 9

cut

cut

cut

© by Math Perspectives

cut

7 +0 ___	6 +1 ___	5 +2 ___	4 +3 ___	3 +4 ___	2 +5 ___	1 +6 ___	0 +7 ___	8 +0 ___
7 +1 ___	6 +2 ___	5 +3 ___	4 +4 ___	3 +5 ___	2 +6 ___	1 +7 ___	0 +8 ___	9 +0 ___
8 +1 ___	7 +2 ___	6 +3 ___	5 +4 ___	4 +5 ___	3 +6 ___	2 +7 ___	1 +8 ___	0 +9 ___

cut

cut

cut

© by Math Perspectives

cut

10 + 0	2 + 8	4 + 6
9 + 1	1 + 9	3 + 7
8 + 2	0 + 10	2 + 8
7 + 3	10 + 0	1 + 9
6 + 4	9 + 1	0 + 10
5 + 5	2 + 8	8 + 2
4 + 6	7 + 3	4 + 6
3 + 7	6 + 4	1 + 9
7 + 3	5 + 5	3 + 7

cut

© by Math Perspectives

cut

1 - 0	4 - 0	5 - 4
2 - 0	4 - 1	5 - 5
2 - 1	4 - 2	6 - 1
1 - 1	4 - 3	6 - 2
2 - 2	4 - 4	6 - 3
3 - 0	5 - 0	6 - 4
3 - 1	5 - 1	6 - 5
3 - 2	5 - 2	6 - 6
3 - 3	5 - 3	6 - 0

cut

cut

cut

© by Math Perspectives

cut

1 -0	1 -1	2 -0	2 -1	2 -2	3 -0	3 -1	3 -2	3 -3
4 -0	4 -1	4 -2	4 -3	4 -4	5 -0	5 -1	5 -2	5 -3
5 -4	5 -5	6 -0	6 -1	6 -2	6 -3	6 -4	6 -5	6 -6

cut

cut

cut

© by Math Perspectives

cut

7 - 0	8 - 1	9 - 1
7 - 1	8 - 2	9 - 2
7 - 2	8 - 3	9 - 3
7 - 3	8 - 4	9 - 4
7 - 4	8 - 5	9 - 5
7 - 5	8 - 6	9 - 6
7 - 6	8 - 7	9 - 7
7 - 7	8 - 8	9 - 8
8 - 0	9 - 0	9 - 9

cut

cut

cut

© by Math Perspectives

cut

7 -0	8 -1	9 -1	8 -2	8 -3	8 -6	9 -0	9 -4	9 -7
7 -1	7 -3	7 -5	7 -7	8 -4	8 -7	9 -2	9 -5	9 -8
7 -2	7 -4	7 -6	8 -0	8 -5	8 -8	9 -3	9 -6	9 -9

cut

© by Math Perspectives

cut

10 - 0	10 - 9	10 -7
10 - 1	10 - 10	10 -8
10 - 2	10 -0	10 -9
10 - 3	10 -1	10 -10
10 - 4	10 -2	10 -2
10 - 5	10 -3	10 -4
10 - 6	10 -4	10 -6
10 - 7	10 -5	10 -7
10 - 8	10 -6	10 -8

cut

© by Math Perspectives

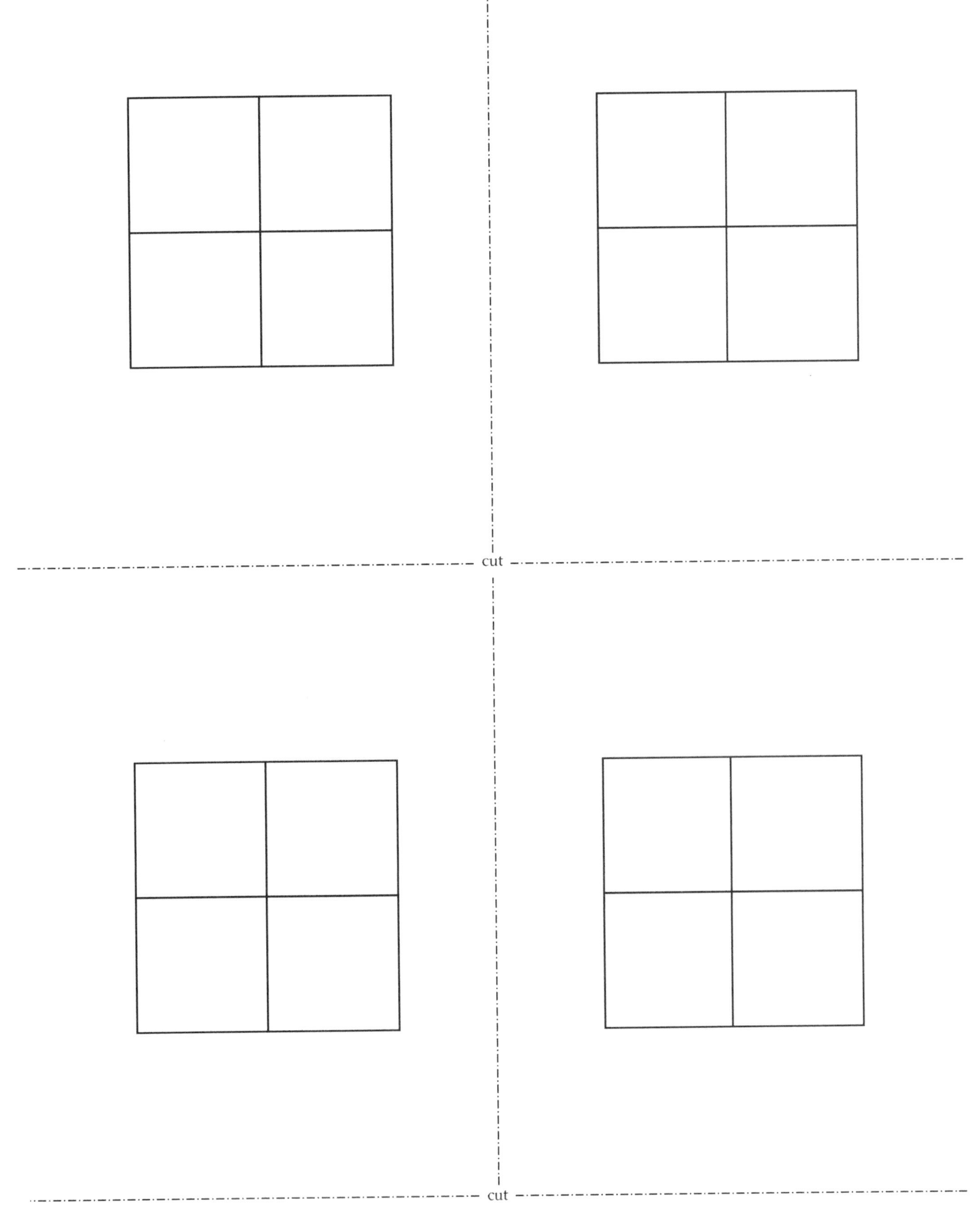

© by Math Perspectives

cut

cut

© by Math Perspectives

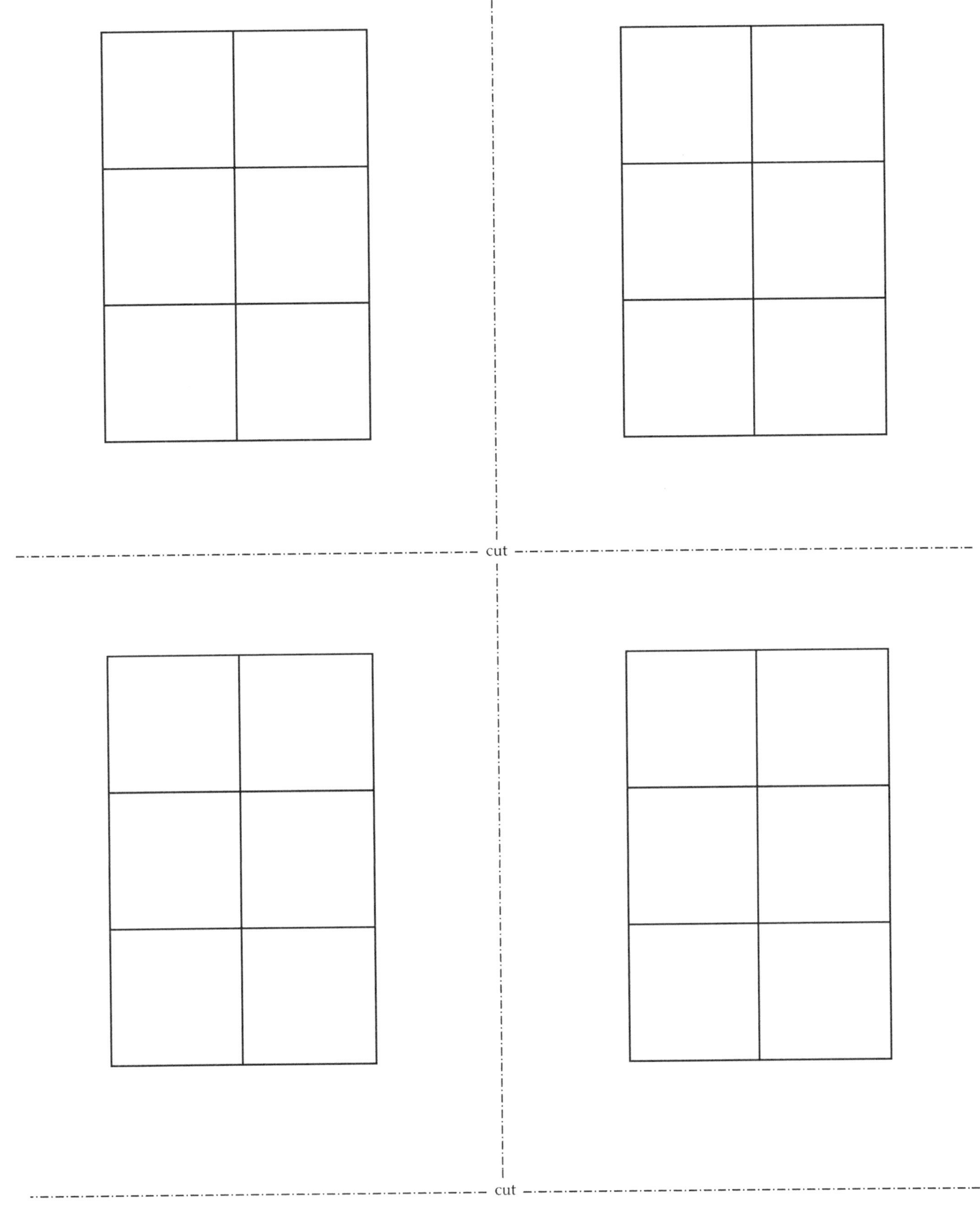

© by Math Perspectives

cut

cut

© by Math Perspectives

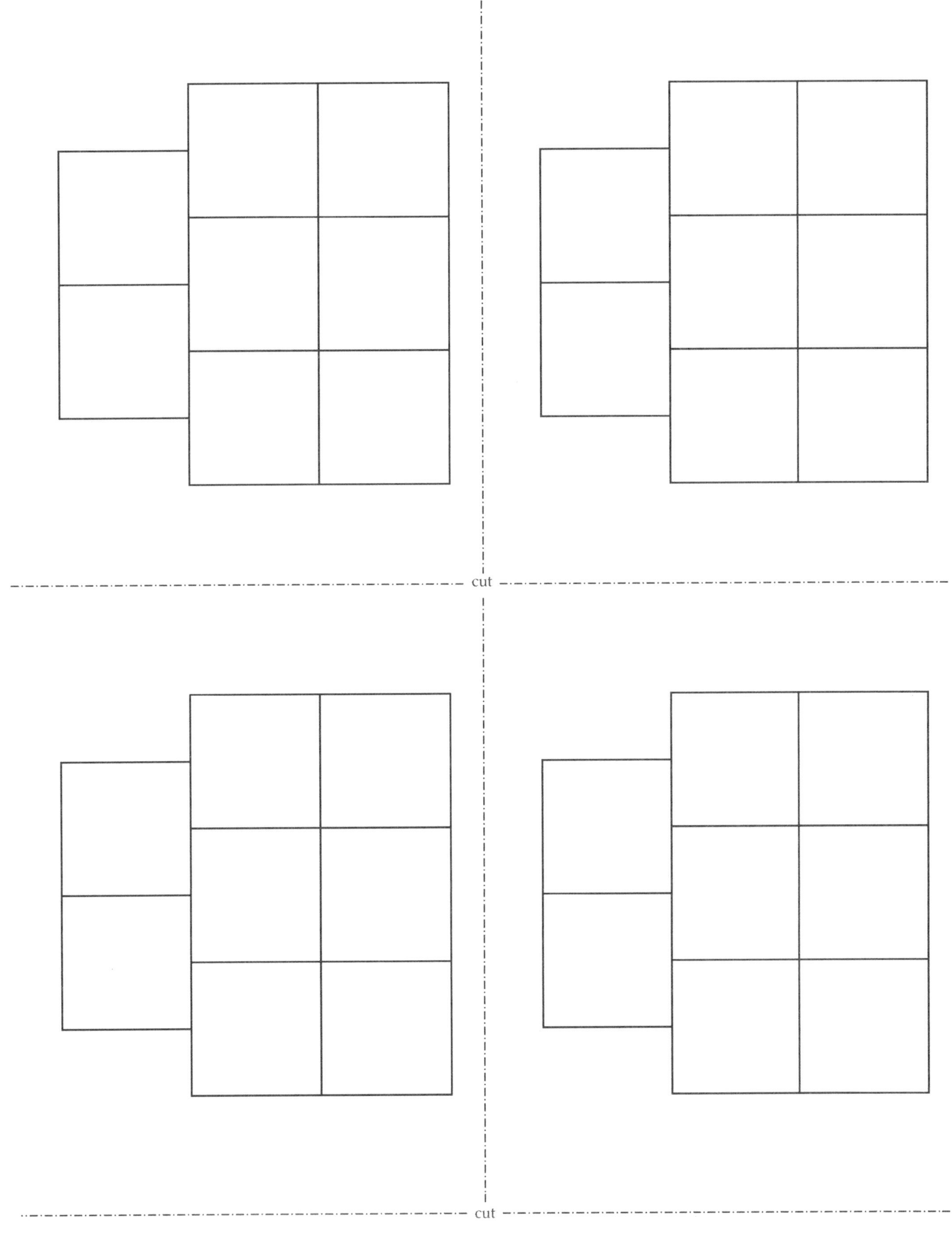

© by Math Perspectives

cut

cut

© by Math Perspectives

cut

cut

cut

cut

© by Math Perspectives

cut

cut

© by Math Perspectives

© by Math Perspectives

© by Math Perspectives

00	01	02	03	04	05	06	07	08	09
10	11	12	13	14	15	16	17	18	19
20	21	22	23	24	25	26	27	28	29
30	31	32	33	34	35	36	37	38	39
40	41	42	43	44	45	46	47	48	49
50	51	52	53	54	55	56	57	58	59
60	61	62	63	64	65	66	67	68	69
70	71	72	73	74	75	76	77	78	79
80	81	82	83	84	85	86	87	88	89
90	91	92	93	94	95	96	97	98	99

© by Math Perspectives

00	01	02	03	04	05	06	07	08	09
10	11	12	13	14	15	16	17	18	19
20	21	22	23	24	25	26	27	28	29
30	31	32	33	34	35	36	37	38	39
40	41	42	43	44	45	46	47	48	49
50	51	52	53	54	55	56	57	58	59
60	61	62	63	64	65	66	67	68	69
70	71	72	73	74	75	76	77	78	79
80	81	82	83	84	85	86	87	88	89
90	91	92	93	94	95	96	97	98	99

00	01	02	03	04	05	06	07	08	09
10	11	12	13	14	15	16	17	18	19
20	21	22	23	24	25	26	27	28	29
30	31	32	33	34	35	36	37	38	39
40	41	42	43	44	45	46	47	48	49
50	51	52	53	54	55	56	57	58	59
60	61	62	63	64	65	66	67	68	69
70	71	72	73	74	75	76	77	78	79
80	81	82	83	84	85	86	87	88	89
90	91	92	93	94	95	96	97	98	99

cut

00	01	02	03	04	05	06	07	08	09
10	11	12	13	14	15	16	17	18	19
20	21	22	23	24	25	26	27	28	29
30	31	32	33	34	35	36	37	38	39
40	41	42	43	44	45	46	47	48	49
50	51	52	53	54	55	56	57	58	59
60	61	62	63	64	65	66	67	68	69
70	71	72	73	74	75	76	77	78	79
80	81	82	83	84	85	86	87	88	89
90	91	92	93	94	95	96	97	98	99

00	01	02	03	04	05	06	07	08	09
10	11	12	13	14	15	16	17	18	19
20	21	22	23	24	25	26	27	28	29
30	31	32	33	34	35	36	37	38	39
40	41	42	43	44	45	46	47	48	49
50	51	52	53	54	55	56	57	58	59
60	61	62	63	64	65	66	67	68	69
70	71	72	73	74	75	76	77	78	79
80	81	82	83	84	85	86	87	88	89
90	91	92	93	94	95	96	97	98	99

cut

© by Math Perspectives

Orange

11 12 21 22 16 17

26 27 51 52 61 62

56 57 66 67

Blue

33 34 35 43 44 45

Green

54 64 74 84 94 85

76 93 82

© by Math Perspectives

Green

4	5	13	14	15	16
23	24	25	26	32	33
34	35	36	37	42	43
44	45	46	47	51	52
53	54	55	56	57	58
61	62	63	64	65	66
67	68				

Brown

74	75	84	85	94	95

© by Math Perspectives

Yellow

00

Brown

10 20 30 40 50 60
70 80 90

Blue

11 12 21 22

Red

13 14 15 16 17 18
19 31 32 33 34 35
36 37 38 39 51 52
53 54 55 56 57 58
59

White

23 24 25 26 27 28
29 41 42 43 44 45
46 47 48 49

© by Math Perspectives

Blue

80 81 82 83 84 85
86 87 88 89 90 91
92 93 94 95 96 97
98 99

Brown

40 41 42 47 48 49
51 52 53 54 55 56
57 58 62 63 64 65
66 67 73 74 75 76

Red

4 14 24 34 44

Orange

5 6 7 8 15 16
17 18

© by Math Perspectives

Blue

10	20	30	40	50	60
70	80	21	31	41	51
61	71	32	42	52	62
43	53	18	28	38	48
58	68	78	88	27	37
47	57	67	77	36	46
56	66	45	55		

Red

2	13	15	6	24	34
44	54	64	74	84	94

© by Math Perspectives

Black

11	12	20	21	22	30
31	32	43	44	45	46
47	48	53	54	55	56
57	59	63	64	65	66
67	72	73	77	78	81
82	83	89	90	91	

© by Math Perspectives

Green

35	36	44	45	46	47
53	54	55	56	57	58
40	41	50	51	62	63
64	65	66	67	68	69
72	79				

© by Math Perspectives

Red

4	13	14	15	22	23
24	25	26	31	32	33
34	35	36	37	40	41
42	43	44	45	46	47
48	50	52	54	56	58

Black

64	74	84	94	82	92
93					

© by Math Perspectives

Name ____________________

I worked with ____________________

	My guess:		I found out:	
	tens	ones	tens	ones

© by Math Perspectives

Name ______________________________

I worked with ______________________________

	My guess:			I found out:		
	hundreds	tens	ones	hundreds	tens	ones

© by Math Perspectives

Name ____________________

How Many More?

I worked with

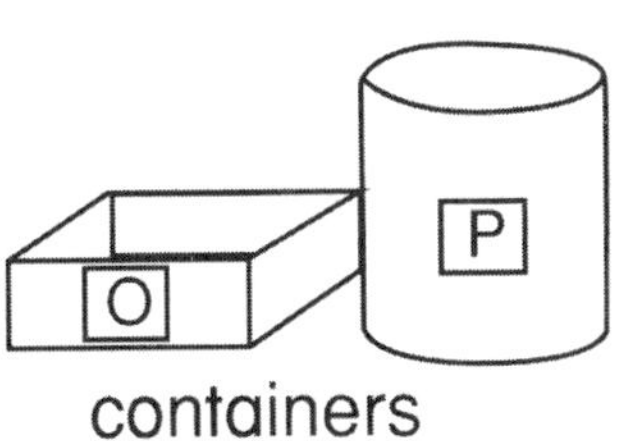

containers

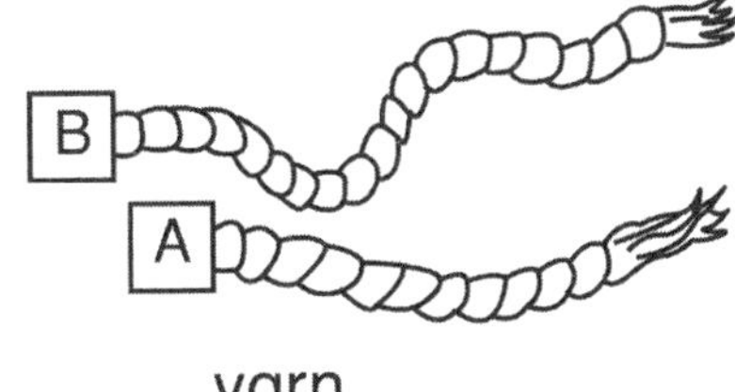

yarn

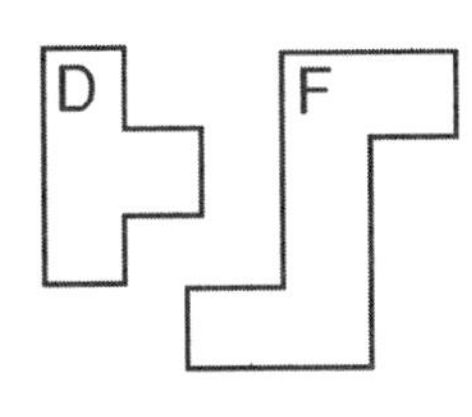

paper shapes

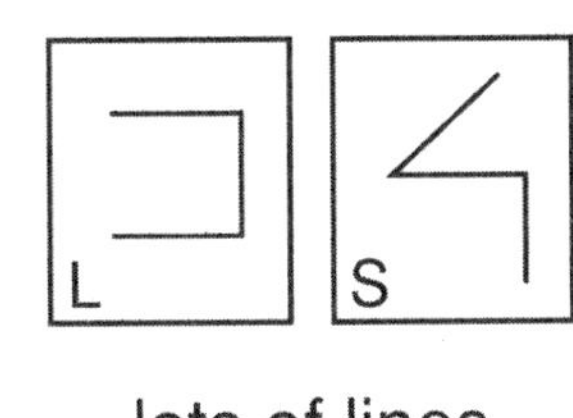

lots of lines

I compared ☐ and ☐.

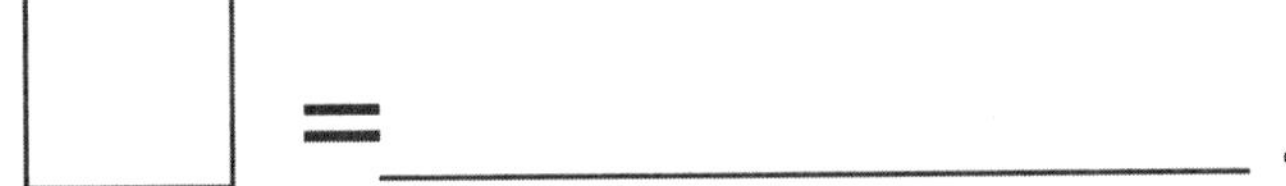
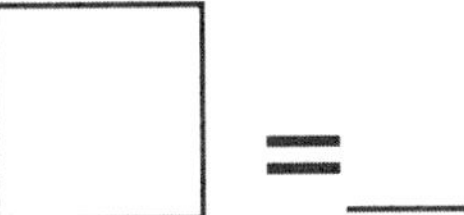
☐ = ____________________. ☐ = ____________________.

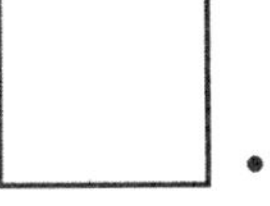
☐ is ____________________ more than ☐.

- - - - - - - - - - cut - - - - - - - - - -

Name ____________________

How Many More?

I worked with

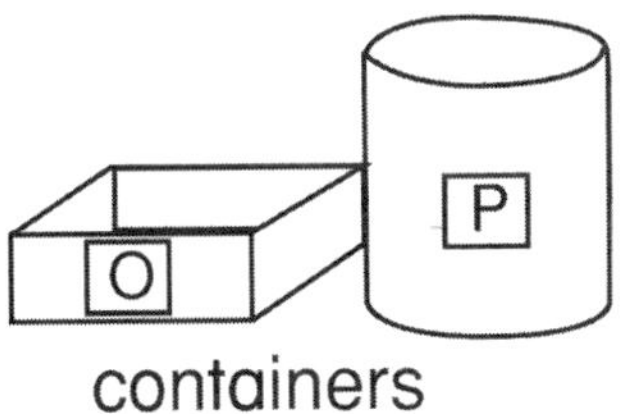

containers

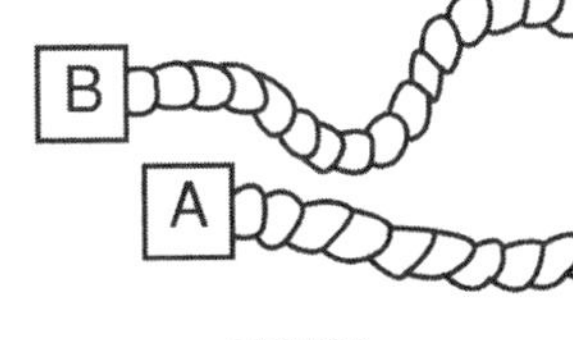
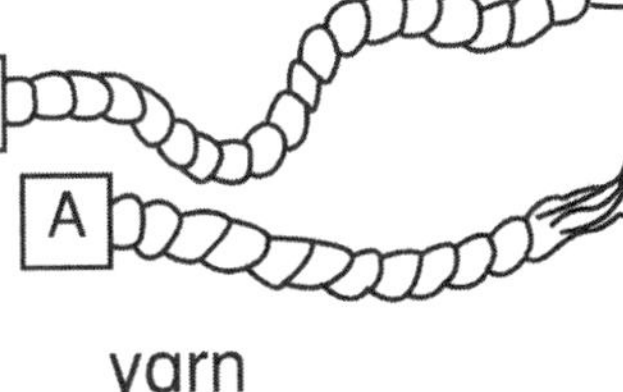

yarn

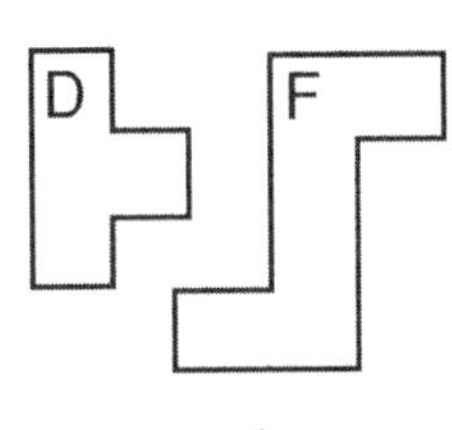

paper shapes

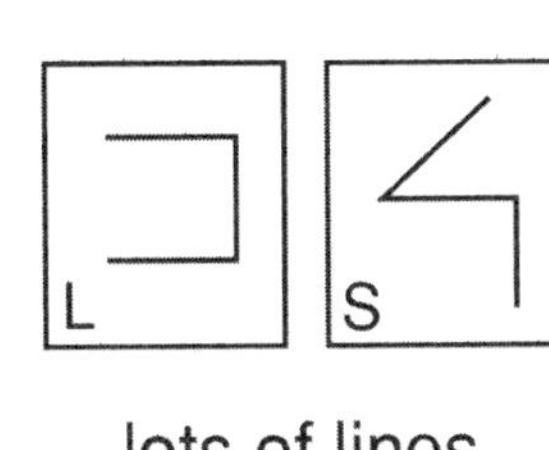

lots of lines

I compared ☐ and ☐.

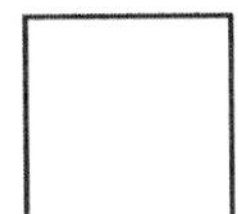
☐ = ____________________. ☐ = ____________________.

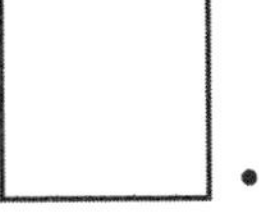
☐ is ____________________ more than ☐.

© by Math Perspectives

Name ______________________

Measuring Things

______________________ = ☐ . ______________________ = ☐ .

______________________ is ☐ longer than ______________________ .

cut

Name ______________________

Measuring Things

______________________ = ☐ . ______________________ = ☐ .

______________________ is ☐ longer than ______________________ .

cut

Name ______________________

Measuring Things

______________________ = ☐ . ______________________ = ☐ .

______________________ is ☐ longer than ______________________ .

cut

Name ______________________

Measuring Things

______________________ = ☐ . ______________________ = ☐ .

______________________ is ☐ longer than ______________________ .

cut

© by Math Perspectives

Name ________________________________

I am________ cubes long.

longer than me	shorter than me
________________	________________
________________	________________
________________	________________
________________	________________
________________	________________
________________	________________
________________	________________

© by Math Perspectives

Name

Measuring Myself

 is ______________________ long.

 is ______________________ long.

 is ______________________ longer than  .

cut

Name

Measuring Myself

 is ______________________ long.

 is ______________________ long.

 is ______________________ longer than  .

© by Math Perspectives

Name ____________________

I made a trail from the

____

to the ____________________.

I guessed ____________________.

I used ____________________.

- - - - - - - - - - cut - - - - - - - - - -

Name ____________________

I made a trail from the

____

to the ____________________.

I guessed ____________________.

I used ____________________.

© by Math Perspectives

Name	Name	Name	Name

tens	ones	tens	ones	tens	ones	tens	ones
1	7	2	6	3	5	4	0
+		+		+		+	
2	2	1	4	2	1	2	4
+		+		+		+	
3	1	3	2	1	9	3	0
+		+		+		+	
2	5	3	6	2	7	1	8
+		+		+		+	

cut cut cut

cut

© by Math Perspectives

Name

tens	ones
8	6
−	
5	9
−	
6	2
−	
7	3
−	

cut

Name

tens	ones
6	0
−	
6	8
−	
7	5
−	
8	1
−	

cut

Name

tens	ones
8	4
−	
8	0
−	
6	7
−	
7	2
−	

cut

Name

tens	ones
8	5
−	
7	4
−	
9	0
−	
6	3
−	

cut

© by Math Perspectives

Name

How many? (cups, rows, groups, stacks)	How many in each?	How many altogether?

© by Math Perspectives

© by Math Perspectives

Name

7 × ____ = ____

6 × ____ = ____

2 × ____ = ____

3 × ____ = ____

4 × ____ = ____

cut

Name

6 × ____ = ____

3 × ____ = ____

2 × ____ = ____

7 × ____ = ____

5 × ____ = ____

cut

Name

4 × ____ = ____

2 × ____ = ____

3 × ____ = ____

6 × ____ = ____

8 × ____ = ____

cut

cut

1 × 4	1 × 6	1 × 7
6 × 1	5 × 1	2 × 1
0 × 3	0 × 7	0 × 4
2 × 0	3 × 0	6 × 0
2 × 2	2 × 3	2 × 4
2 × 5	2 × 6	2 × 7
3 × 2	3 × 3	3 × 4
4 × 3	3 × 5	3 × 6
4 × 2	4 × 5	4 × 6

cut

© by Math Perspectives

cut

5 × 2	5 × 3	5 × 4
3 × 7	3 × 8	2 × 8
5 × 5	5 × 6	6 × 2
6 × 3	6 × 4	6 × 5
1 × 2	1 × 3	1 × 5
2 × 9	4 × 4	4 × 7
7 × 2	7 × 3	8 × 2
8 × 3	9 × 2	9 × 3
1 × 9	9 × 1	0 × 9

cut

cut

cut

© by Math Perspectives

Name ______________________

______ × 0 = ______

______ × 1 = ______

______ × 2 = ______

______ × 3 = ______

______ × 4 = ______

______ × 5 = ______

______ × 6 = ______

______ × 7 = ______

______ × 8 = ______

______ × 9 = ______

cut

Name ______________________

______ × 0 = ______

______ × 1 = ______

______ × 2 = ______

______ × 3 = ______

______ × 4 = ______

______ × 5 = ______

______ × 6 = ______

______ × 7 = ______

______ × 8 = ______

______ × 9 = ______

© by Math Perspectives

Name ______________________ | Name ______________________

$0 \times$ ______ = ______

$1 \times$ ______ = ______

$2 \times$ ______ = ______

$3 \times$ ______ = ______

$4 \times$ ______ = ______

$5 \times$ ______ = ______

$6 \times$ ______ = ______

$7 \times$ ______ = ______

$8 \times$ ______ = ______

$9 \times$ ______ = ______

cut

$0 \times$ ______ = ______

$1 \times$ ______ = ______

$2 \times$ ______ = ______

$3 \times$ ______ = ______

$4 \times$ ______ = ______

$5 \times$ ______ = ______

$6 \times$ ______ = ______

$7 \times$ ______ = ______

$8 \times$ ______ = ______

$9 \times$ ______ = ______

© by Math Perspectives

Name ____________________

$\overline{)\,31}$

$\overline{)\,16}$

$\overline{)\,29}$

$\overline{)\,20}$

cut

Name ____________________

$\overline{)\,35}$

$\overline{)\,14}$

$\overline{)\,10}$

$\overline{)\,24}$

cut

Name ____________________

$\overline{)\,32}$

$\overline{)\,36}$

$\overline{)\,12}$

$\overline{)\,18}$

© by Math Perspectives

© by Math Perspectives

Name

$25 \div$ ____ = ____

$15 \div$ ____ = ____

$32 \div$ ____ = ____

$18 \div$ ____ = ____

cut

Name

$12 \div$ ____ = ____

$23 \div$ ____ = ____

$37 \div$ ____ = ____

$10 \div$ ____ = ____

cut

Name

$17 \div$ ____ = ____

$9 \div$ ____ = ____

$27 \div$ ____ = ____

$31 \div$ ____ = ____

$6 \div 3$	$6 \div 2$	$6 \div 1$
$6 \div 4$	$7 \div 2$	$7 \div 3$
$8 \div 2$	$8 \div 3$	$8 \div 4$
$9 \div 3$	$9 \div 2$	$9 \div 4$
$12 \div 3$	$12 \div 4$	$12 \div 6$
$12 \div 5$	$14 \div 7$	$14 \div 2$
$15 \div 3$	$15 \div 2$	$15 \div 5$
$15 \div 4$	$20 \div 4$	$20 \div 3$
$20 \div 5$	$20 \div 2$	$25 \div 4$

© by Math Perspectives

Professional Development Support for *Developing Number Concepts* Teachers

For information on the Mathematical Perspectives Courses and Workshops developed by Kathy Richardson to support the teaching approach in the *Developing Number Concepts* series and Planning Guide, contact:

Math Perspectives
Kathy Richardson and Associates
P.O. Box 29418
Bellingham, WA 98228–9418
Phone: 360–715–2782
Fax: 360–715–2783

Notes

Notes

Notes